Learning for Action

Learning for Action

A Short Definitive Account of Soft Systems Methodology and its use for Practitioners, Teachers and Students

Peter Checkland

John Poulter

John Wiley & Sons, Ltd

Telephone (+44) 1243 779777

Email (for orders and customer service enquiries): cs-books@wiley.co.uk
Visit our Home Page on www.wiley.com

Reprinted December 2007, January 2009

Other Wiley Editorial Offices

John Wiley & Sons Inc., 111 River Street, Hoboken, NJ 07030, USA

Jossey-Bass, 989 Market Street, San Francisco, CA 94103-1741, USA

Wiley-VCH Verlag GmbH, Boschstr. 12, D-69469 Weinheim, Germany

John Wiley & Sons Australia Ltd, 42 McDougall Street, Milton, Queensland 4064, Australia

John Wiley & Sons (Asia) Pte Ltd, 2 Clementi Loop #02-01, Jin Xing Distripark, Singapore 129809

John Wiley & Sons Canada Ltd, 22 Worcester Road, Etobicoke, Ontario, Canada M9W 1L1

Wiley also publishes its books in a variety of electronic formats. Some content that appears in print may not be available in electronic books.

Library of Congress Cataloging in Publication Data

Checkland, Peter.
Learning for action: a short definitive account of soft systems methodology and its use for practitioner, teachers, and students/Peter Checkland, John Poulter.
p. cm.
Includes bibliographical references and index.
ISBN-13 978-0-470-02554-3 (pbk.: acid-free paper)
1. System theory. 2. Problem solving. 3. System analysis. I. Poulter, John. II. Title.
Q295.C448 2006
658.4′032—dc22 2006010828

British Library Cataloguing in Publication Data

A catalogue record for this book is available from the British Library

ISBN-13 978-0-470-02554-3 (PB)

Typeset in 10/16pt Kunstler by Integra Software Services Pvt. Ltd, Pondicherry, India
Printed and bound in Great Britain by TJ International Ltd, Padstow, Cornwall

To the memory of Glen Checkland (1930–1990)

PBC

To Peter Checkland and the humility that comes from great understanding

JP

Contents

Foreword

This is a very welcome book for anyone concerned with turning the tide against the rise and rise of systemic failure!! It will deliver rewards whether you are a health worker, an ICT specialist, a public servant, a politician, a policy advisor, a student, an environmental activist, a researcher, a teacher, a parent, in fact anyone involved in managing... and that is most of us!

This book is welcome because it is accessible and practical, built on success in managing change and improvement in organizations and other complex situations as well as rigorous scholarship. SSM, as described in this book, is not a tool or technique to be used occasionally but a way to think and act every day. It is an antidote to the instrumental and naive ways SSM has been taught to a generation of MBA students. It is also a provocative challenge to those academics and practitioners stuck in a 1960s view of systems scholarship.

One of Checkland and Poulter's main messages is that it is only by taking part in SSM practice that you will really understand and enjoy the benefits... and what they have successfully done is to write the ideal companion for getting involved in improving messy, complex situations through the use of SSM – a proven and practical way to engage in reflective practice.

Above all else, this is a book which can enable you to make SSM your own – to adopt and adapt for use 'in any situation in which people are intent on taking action to improve it'!

Ray Ison
Professor of Systems
Director of the Open Systems Research Group
Systems Department
The Open University
Walton Hall
Milton Keynes
MK7 6AA

Preface

Our aim is a very simple one: to provide an account of Soft Systems Methodology (SSM) that is short, definitive and will enable anyone interested to learn about it, to teach it or to start using it in real situations.

The first account of the approach to tackling real-world problem situations that became known as Soft Systems Methodology was published in 1972. Over the years since then the story of its development has been told in detail in several different ways: in four books, in several chapters in various other books, and in a large number of papers, mainly in journals read only by scholars. These describe the evolution of the approach in a 30-year programme of practical research based at Lancaster University. This research was carried out in problematical situations in organizations outside the university. It started out by taking the approach known as Systems Engineering, as developed in the Bell Telephone Laboratories, and testing whether or not it could be used outside the technological situations for which it was originally developed. Bell had examined case histories of their own technical developments and from them generalized a procedure for making any such development. Our initial question was: Could this approach perhaps also be applied in *management* problem situations? In the event, the pattern of activity found in Systems Engineering – namely, precisely define a need and then engineer a system to meet that need using various techniques – was simply not rich enough to deal with the buzzing complexity and confusion of management situations. Thus, technological development is usually carried out in response to a given perceived need – such as, for example, a requirement for a new telecommunication system, or a better mousetrap – whereas, in management situations, defining the need to be met is itself always part of the problem. Management always entails What to do? as well as How to do it? Given this, the Lancaster research saw the emergence of a radical alternative to Systems Engineering, namely the new approach which became known as SSM.

Having been honed in several hundred projects within the programme of action research, SSM is now a mature and well-tested approach. This means that to use it you don't have to know about its origins, the course of its development, its underlying theory or its intellectual context. All of these are described in detail in the existing literature, if required. None of those aspects will be covered here, however. We kept in mind two questions:

- What is SSM?
- How can it be used?

Thus our account focuses on the approach itself, with only a brief summary of its underlying theory in an Appendix which indicates where more detailed discussion may be found.

Apart from people who may wish to start using SSM, we have two other relevant audiences in mind. There is now a considerable 'secondary literature' about SSM, but unfortunately much of it is of rather poor quality, littered with misunderstandings and inaccuracies. In her research on the emergence of SSM Sue Holwell found the classic example in a 1995 book about information systems: a passage of fewer than 200 words which contains nearly twenty prime errors! Given this situation we provide here a reliable account of SSM which we hope will be helpful to those who teach courses dealing with the approach in colleges, universities and organizations.

Equally, we intend that the account will help students learning about SSM to grasp its fundamentals fairly quickly and – we trust – relatively painlessly. We mention the possible 'pain' because the fact is that these errors in much of the secondary literature occur as a result of a failure to grasp the radical nature of SSM compared with Systems Engineering and similar approaches. In order to take on board the richness of management (i.e. human) problem situations, SSM had to develop new ways of thinking about the complexity of real-life situations. These took the approach a good step away from the kind of thinking required to find a way to achieve an objective which can be taken as given from the start. This shift in thinking means that for many people understanding SSM involves some rearranging of their mental furniture. This is never an entirely soothing experience! Given this, and in spite of it, our aim is to make SSM seem 'obvious', so that gaining an understanding of it is made as comfortable a process as possible.

Before the material of the book itself, the Preamble offers a summary for those with only ten minutes to spare!

We are able to write this concise account of SSM thanks to the contributions of the very large number of people who enriched its long development. They are too many to name, but we are grateful to them all. We can name here, invidiously perhaps, only a group of reflective practitioners with whom we have had the good fortune to collaborate, and who have made valuable contributions towards advancing SSM: Paulo Nunes de Abreu, the late Ron Anderton, Chris Atkinson, David Brown, Alex Casar, Sheila Challender, Steve Clarke, Elaine Cole, Lynda Davies, Roger Elvin, Paul Forbes, John Hardy, Mike Haynes, Kees van der Heiden, Luc Hoebeke, Sue Holwell, Nimal Jayaratna, Paul Ledington, Jaap Leemhuis, Iain Perring,

Chris Pogson, Sophia Martin, Jim Scholes, David Smyth, Frank Stowell, Ken Uchiyama, Werner Ulrich, Brian Wilson, Mark Winter, the late Peter Wood.

Finally many thanks go to Jenny Seddon for her willing and cheerful professionalism in producing clean copy from a handwritten manuscript.

Preamble

A Ten-Minute Account of Soft Systems Methodology for Very Busy People

- We all live in the midst of a complex interacting flux of changing events and ideas which unrolls through time. We call it 'everyday life', both personal and professional. Within that flux we frequently see situations which cause us to think: 'Something needs to be done about this, it needs to be improved.' Think of these as 'problematical situations', avoiding the word 'problem' since this implies 'solution', which eliminates the problem for ever. Real life is more complex than that!
- Soft Systems Methodology (SSM) is an organized way of tackling perceived problematical (social) situations. It is action-oriented. It organizes thinking about such situations so that action to bring about improvement can be taken.
- The complexity of problematical situations in real life stems from the fact that not only are they never static, they also contain multiple interacting perceptions of 'reality'. This comes about because different people have different taken-as-given (and often unexamined) assumptions about the world. This causes them to see it in a particular way. One person's 'terrorism' is another's 'freedom fighting'; one person sees a prison in terms of punishment, another sees it as seeking rehabilitation. These people have different *worldviews*. Tackling problematical situations has

to accept this, and has to pitch analysis at a level that allows worldviews to be surfaced and examined. For many people worldviews are relatively fixed; but they can change over time. Sometimes a dramatic event can change them very quickly.

- All problematical situations, as well as containing different worldviews, have a second important characteristic. They always contain people who are trying to act *purposefully*, with intention, not simply acting by instinct or randomly thrashing about – though there is always plenty of that too in human affairs.
- The previous two points – the existence of conflicting worldviews and the ubiquity of would-be purposeful action – lead the way to tackling problematical situations. They underpin the SSM approach, a process of inquiry which, through social learning, works its way to taking 'action to improve'. Its shape is as follows:

  1. Find out about both the problematical situation and the characteristics of the intervention to improve it: the issues, the prevailing culture and the disposition of power within the overall situation (its politics). Ways of doing these things are provided.
  2. From the finding out, decide upon some relevant purposeful activities, relevant that is to exploring the situation deeply, and remembering that the ultimate aim is to define and take 'action to improve'. Express these relevant purposeful activities as activity *models*, each made to encapsulate a declared worldview, the model being a cluster of linked activities which together make up a purposeful whole. (For example, one model could express in terms of activities the notion 'prison' as if it were only 'a punishment system', another could express it as 'a rehabilitation system'.) Such models never describe the real world, simply because they are based on one pure worldview. They are devices, or tools, to explore it in an organized

Part one
The nature of SSM

1
A Skeleton Account of SSM

What is SSM?

The aim of the work which led to the development of Soft Systems Methodology (SSM) was to find a better way of dealing with a kind of situation we continually find ourselves facing in everyday life: a situation about which we have the feeling that 'something needs to be done about this'. We shall call such situations 'problematical', rather than describing them as 'problem situations', since they may not present a well-defined 'problem' to be 'solved' out of existence – everyday life is more complex than that! A company might feel that it needs to stimulate sales, perhaps by introducing a new product; or should they bid for the equity of a smaller rival? A university may feel that its student intake is too biased towards students from middle-class homes. What are the implications of changing that? A government may struggle to define legislation which would increase the feeling of security on the streets, given the threat of terrorism, without diminishing civil liberties. A local council may be receiving complaints that the delivery of its services is not sufficiently 'citizen-friendly'. What should it do? A head teacher may wonder how to decide whether to take on the responsibility for providing school meals (the school benefiting from any surplus generated) or to leave that function to the local education authority. An individual may develop a sense of unease about the future viability of the firm he or she works for, and wonder whether to look for a job elsewhere. All these are 'problematical situations'. They could be tackled in various ways: by appealing to previous experience; intuitively; by randomly thrashing about (never a shortage of that in human situations); by responding emotionally; or they could be addressed by using SSM.

So what is it? It is an organized, flexible process for dealing with situations which someone sees as problematical, situations which call for action to be taken to improve them, to make them more acceptable, less full of tensions and unanswered questions. The 'process' referred to is an organized process of thinking your way to taking sensible 'action to improve' the situation; and, finally, it is a process based on a particular body of ideas, namely *systems* ideas.

That these ideas have proved themselves to be useful in dealing with the complexity of the social world is hardly surprising. Social situations are always complex due to multiple interactions between different elements in a problematical situation as a whole, and systems ideas are fundamentally concerned with the *interactions* between parts of a whole. So it is systems ideas which help to structure the thinking. (However, the way systems ideas are used within SSM is fundamentally different from the way they inform the various earlier systems approaches developed in the 1950s and 1960s, as we shall see below.)

In order to ensure that the previous two paragraphs are clear, we need to unpack them somewhat, and say a little more about the crucial elements within them, if this chapter is to fulfil its aim of presenting a broad-brush account of SSM as a whole. Four elements in the paragraphs above will be expanded: 'everyday life and problematical situations'; 'tackling such situations'; a 'flexible process', and 'the use of systems ideas'.

Everyday Life and Problematical Situations

As members of the human tribe we experience everyday life as being quite exceptionally complex. We feel ourselves to be carried along in an onrushing

The Use of Systems Ideas

As stated above, systems ideas concern interaction between parts which make up a whole; also, the complexity of real situations is always to a large extent due to the many interactions between different elements in human situations. So it is not surprising that systems ideas have some relevance to dealing with real-world complexity (though they are only very rarely useful in *describing* that complexity).

The core systems idea or concept is that of an adaptive whole (a 'system') which can survive through time by adapting to changes in its environment. The concept is illustrated in Figure 1.1. A system S receives shocks from

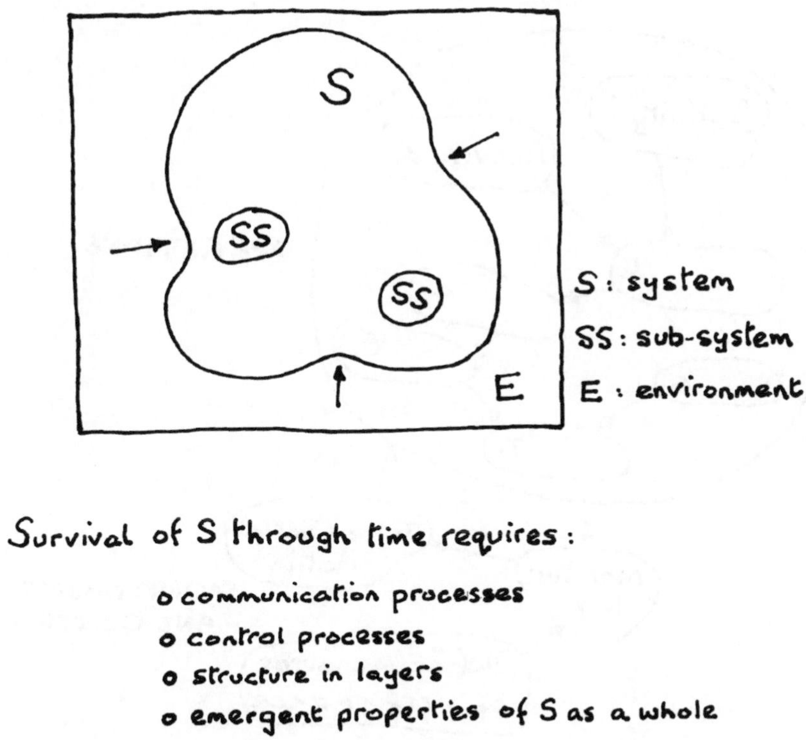

Figure 1.1 The core systems concept: an adaptive whole

its changing environment E. If it is to survive it requires *communication processes* (to know what is going on) and *control processes* (possible adaptive responses to the shocks). Also, the system may contain sub-systems SS, or may itself be seen by a different observer as only a sub-system of some wider system. The idea of a *layered structure* is thus fundamental in systems thinking. Finally, what is said to be a system must have some properties as a single whole, so-called *emergent properties*. (Thus the parts of a bicycle, when assembled correctly, and only then, produce a whole which has the emergent property of being a vehicle, the concept 'vehicle' being meaningful only in relation to the whole.) These four italicized phrases represent the core of systems thinking. So how can it be used here?

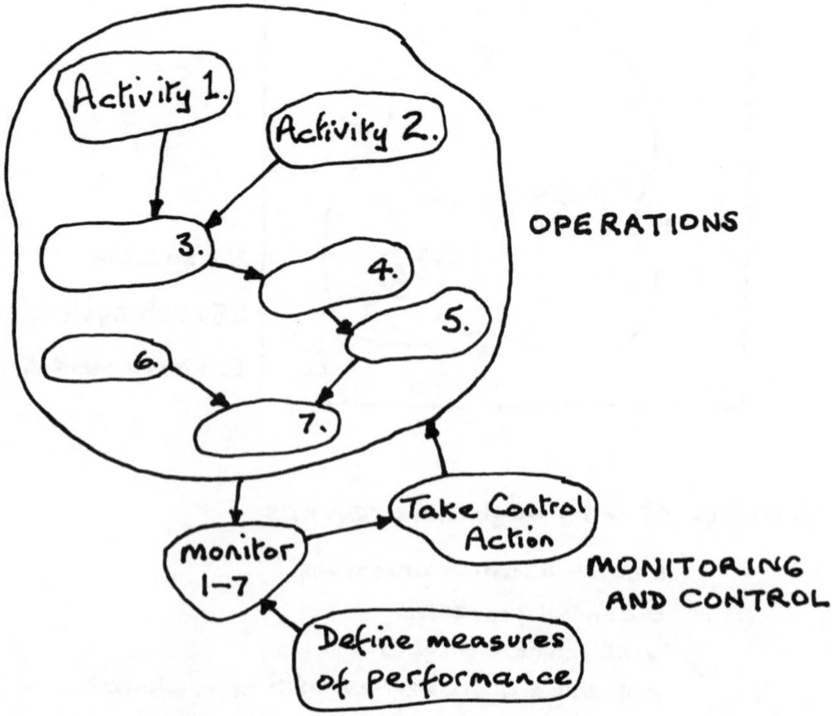

Figure 1.2 The general form of a purposeful activity model

The relevance of this kind of thinking to SSM emerged when it was realized that every single real-world problematical situation, whether in a small firm making wheelbarrows, a multi-national oil company, or in the National Health Service (which employs more than a million people) has one characteristic in common. All such situations contain people trying to act *purposefully*, not simply acting by instinct or splashing about at random. From this observation comes the key idea of *treating purposeful action as a system*. A way of representing purposeful action as a system, i.e. an adaptive whole (in line with Figure 1.1) was invented. Figure 1.2 shows its general form.

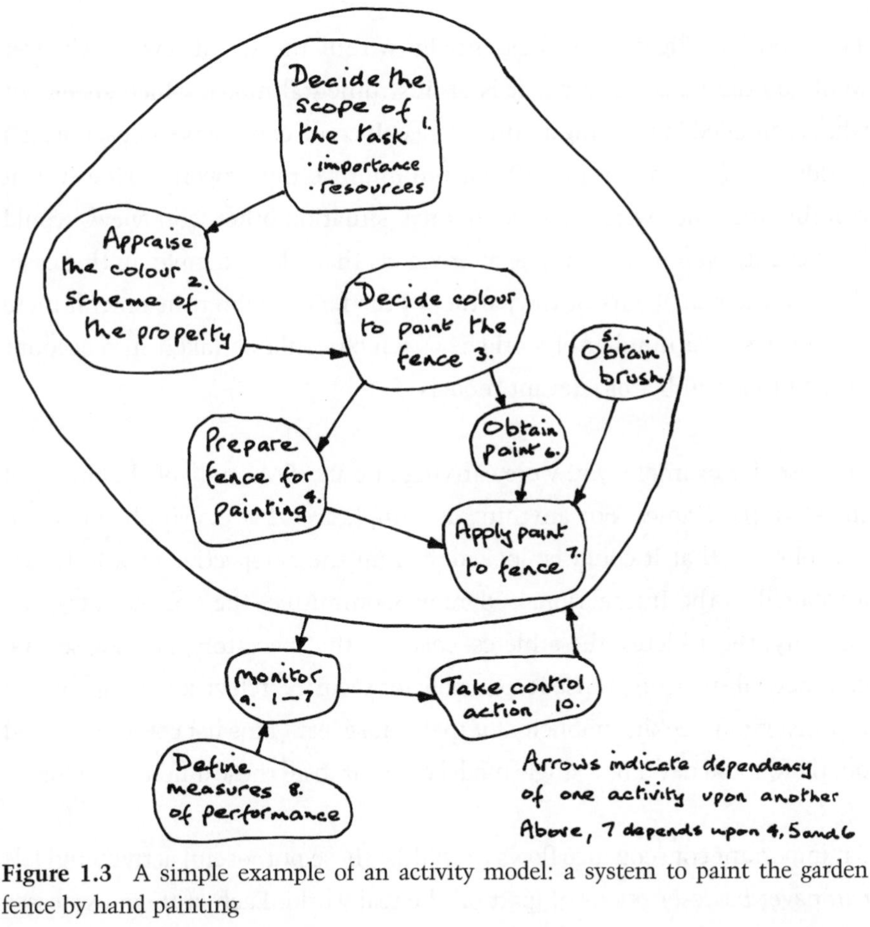

Figure 1.3 A simple example of an activity model: a system to paint the garden fence by hand painting

A logically linked set of activities constitute a whole – its emergent property being its purposefulness. The activities concerned with achieving the purpose (the operations) are monitored against defined measures of performance so that adaptive control action (to make changes) can be taken if necessary.

Figure 1.3 shows a trivial example to illustrate the concept. With regard to Figure 1.2, the 'measure of performance' might be the degree to which fence painting enhances the appearance of the property or, perhaps, 'good' or 'bad' might be defined according to whether or not the neighbours complain about it. This model, then, is a 'purposeful activity model'.

The model in Figure 1.3 is essentially within the worldview of whoever would do the fence painting. It is an instrumental model which spells out what is entailed in painting a garden fence. It could express the householder's worldview: 'I can do useful DIY jobs to improve my property.' However, if painting the fence were an issue in a real situation other worldviews would be relevant, even in an example as trivial as this – for example, in this case, those of the neighbours or the partner of the fence-painter. In general there will always be a number of worldviews which could be taken into account leading to a number of relevant models.

Suppose, for example, you were carrying out an SSM study of the future of the Olympic Games. For anything as complex as this global phenomenon it is obvious that it could be looked at from the perspective of worldviews attributed to the International Olympic Committee, the host country, the host city, the athletes, the athletes' coaches, the spectators, hot dog sellers, commercial sponsors, those responsible for security, television companies, a terrorist group seeking publicity for their cause, etc. This list could go on and on; there could never be a single model relevant to all these different interests.

An important consequence flows from this: these purposeful activity models *can never be descriptions* of (part of) the real world. Each of them expresses *one way of looking at and thinking about* the real situation, and there will be multiple possibilities. So how can such models be made useful? The answer is to see them as *devices* (intellectual devices) which are a source of *good questions to ask about the real situation*, enabling it to be explored richly. For example, we could focus on the differences between a model and the situation, and ask whether we would like activity in the situation to be more, or less, like that in the model. Such questioning organizes and structures a discussion/debate about the real-world situation, the purpose of that discussion being to surface different worldviews and to seek possible ways of changing the problematical situation for the better. This means finding an accommodation, that is to say a version of the situation which different people with different worldviews could nevertheless live with. Given the different worldviews which will always be present in any human situation, this means finding possible changes which meet two criteria simultaneously. They must be arguably *desirable*, given the outcomes of using the models to question the real situation, but must also be culturally *feasible* for these particular people in this particular situation with its unique history and the unique narrative which its participants will have constructed over time in order to make sense of their experience. Figure 1.4 illustrates this.

In summary, then, we have:

- a problematical real-world situation seen as calling for action to improve it;
- models of purposeful activity *relevant* to this situation (not describing it);
- a process of using the models as devices to explore the situation;
- a structured debate about desirable and feasible change.

This gives the bare bones of the process of SSM, whose shape can now be described.

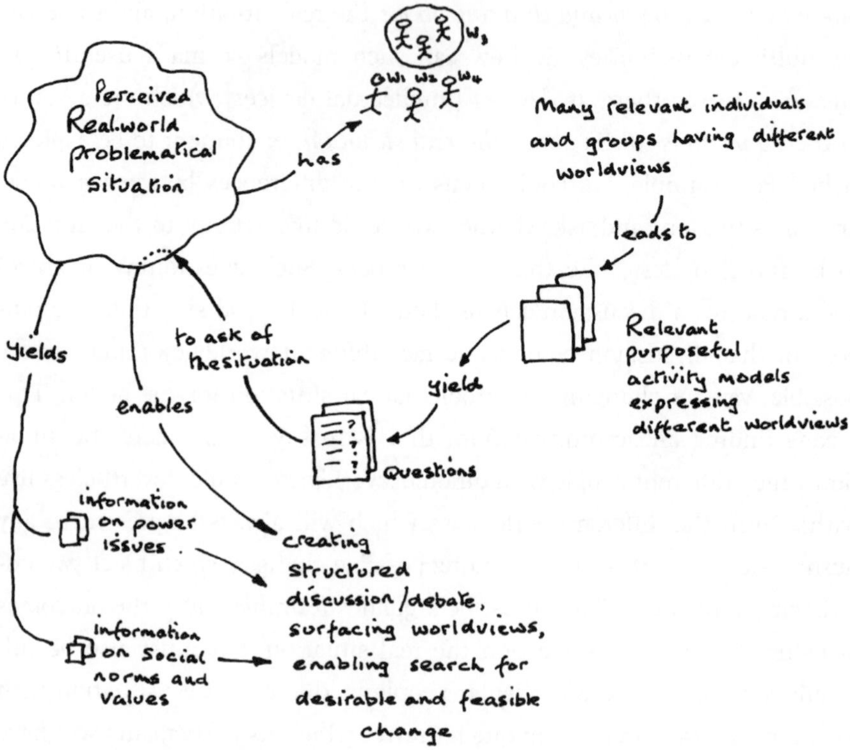

Figure 1.4 SSM's basic process

What is the SSM Process?

The SSM process takes the form of a cycle. It is, properly used, a cycle of learning which goes from finding out about a problematical situation to defining/taking action to improve it. The learning which takes place is social learning for the group undertaking the study, though each individual's learning will be, to a greater or lesser extent, personal to them, given their different experiences of the world, and hence the different worldviews which they will bring to the study. Taking action as a result of the study will of course change the starting situation into a new situation, so that in principle the cycle could begin again (a relevant system then being 'a system to make

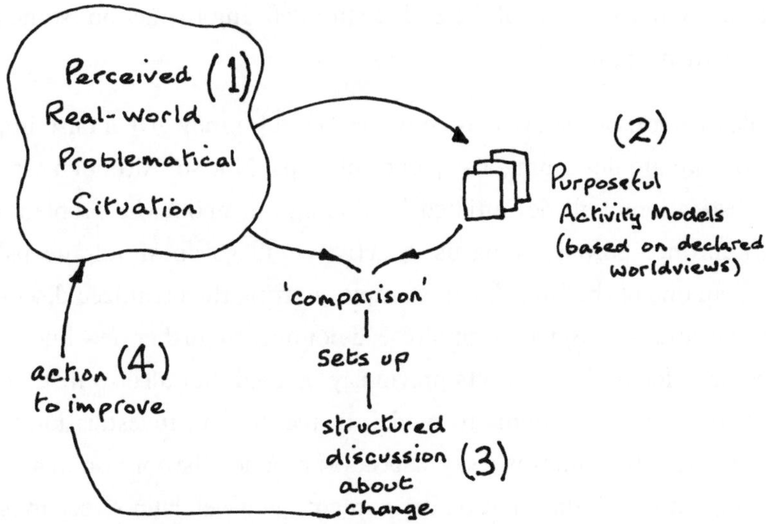

Figure 1.5 The iconic representation of SSM's learning cycle

these changes'). SSM is thus not only a methodology for a specially set-up study or project; it is, more generally, a way of managing any real-world purposeful activity in an ongoing sense.

The SSM cycle is shown in Figure 1.5, which eventually emerged as its classic representation. It contains four different kinds of activity:

1. Finding out about the initial situation which is seen as problematical.
2. Making some purposeful activity models judged to be relevant to the situation; each model, as an intellectual device, being built on the basis of a particular pure worldview.
3. Using the models to question the real situation. This brings structure to a discussion about the situation, the aim of the discussion being to find changes which are both arguably desirable and also culturally feasible in this particular situation.
4. Define/take the action to improve the situation. Since the learning cycle is in principle never-ending it is an arbitrary distinction as to whether the end of a study is taken to be defining the action or actually carrying

it out. Some studies will be ended after defining the action, some after implementing it.

This description of the cycle as activities (1) to (4) may give a false impression that we are describing a sequence of steps. Not so. Although virtually all investigations will be initiated by finding out about the problematical situation, once SSM is being used, activity will go on simultaneously in more than one of the 'steps'. For example, starting the organized discussion about the situation (3) will normally lead not only to further new finding out (1), perhaps focused on aspects previously ignored, but also to further new choices of 'relevant' systems to model. In real life, an investigation which sets out narrowly to improve, say, aspects of product distribution in a manufacturing company's distribution department, may well later sweep in issues concerning, perhaps, communications between production and marketing departments. Figure 1.6 illustrates a typical pattern of activity of the kind which emerges as an investigation digs deeper.

Figure 1.6 shows an on-going 'finding out' activity, three bursts of model building, discussion fed by both the models and the finding out, which itself leads to more finding out and more modelling. The final (fourth) burst of modelling shown here as an example follows from defining the 'action to improve' and would consist of purposeful activity models relevant to carrying out the action agreed.

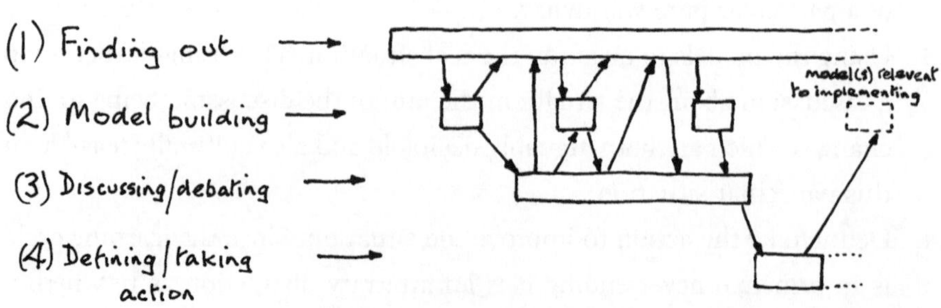

Figure 1.6 A typical pattern of activity during an SSM investigation

Finally, in describing the SSM cycle, we could add (though this is really a point from the end of this book) that as users of SSM become more sophisticated they treat Figure 1.5 not at all as a prescription to be followed, but as a model to make sense of their experience as they mentally negotiate their way through the problematical situation.

What Can SSM be Used for?

The application area for SSM is very broad. This is not due to megalomania on the authors' part. Rather it stems from the wide applicability of two key ideas behind SSM. One of these is to create a process of *learning your way* through problematical situations to 'action to improve' – a very general concept indeed. The other is the idea that you can make sure this learning is organized and structured by using, as a source of questions to ask in the real situation, models (systems models) of purposeful activity. This is because every real-world situation contains people trying to act purposefully, intentionally. It is the sheer generality of purposeful action – the core of being human – that makes the area in which SSM can be used so huge.

In Part Two, the stories of SSM use come from all sizes of company from small firms to large corporations, from organizations in both private and public sectors, including the National Health Service. Chapter 4 describes uses of SSM in the world of information systems and information technology, where it is much used. This derives from the fact that for any purposeful activity model (Figure 1.3 being a noddy example) you can ask of each activity: What information would support doing this activity? And what information would be generated by doing it? Since information is what you get when you attribute meaning to data in a particular context, and meaning attribution depends upon worldview, SSM's strong emphasis on worldview explains its relevance to this field.

In summary, SSM can be used in any human situation which entails thinking about acting purposefully, and is especially useful in any situation in which it is helpful to lift the level of discussion from that of everyday opinions and dogma to that level at which you are asking: What taken-as-given worldview lies behind these assertions of opinion?

Is SSM Mature?

Obviously it is never possible to claim that the development of any approach to human inquiry is 'finished', though some features of any such process may become so taken-as-given as to appear permanent. For example, in the inquiry process of natural science, if you are testing a new drug you give some patients the drug while others receive a placebo. The difference between the group ingesting the drug and the so-called 'control' group taking the placebo tells you what effects the drug produces (given a statistically significant sample size). This pattern would seem to be a permanent feature of scientific experiment. In applied social science, where SSM sits, the situation is less definite. Nevertheless, after hundreds of studies the core processes of SSM do now appear to be well-established, though the application area continues to expand. In the early days each significant study was likely to cause some rethinking of the process itself; but such changes became increasingly rare over the 30-year development period. We now regard it as a mature process.

The most recent addition to the literature about its development describes the use of SSM both in relation to the perceived *content* of the situation in question – SSM (c) – and in relation to the *process* of carrying out the inquiry itself – SSM (p) – (described in Chapter 2). This is in a paper published in 2006. But this is a case of the literature lagging behind practice, as these twin uses of SSM have been recognized and exploited by those developing the approach since the early 1980s.

So SSM is now considered mature enough to justify writing this book.

How was SSM Created?

The classic way of doing research comes from natural science: set up a hypothesis and then test it experimentally. It is not easy to transfer this model of research to the gloriously rich social and human arena, though strenuous efforts to do that have been made over many years. SSM was developed using an alternative model of research, one more suitable for social research at the level of a situation, group or organization, namely 'action research'. In this kind of research you accept the great difficulty of 'scientific' experimental work in human situations, since each human situation is not only unique, but changes through time and exhibits multiple conflicting worldviews. Hence the pattern for the action researcher is to enter a human situation, *take part* in its activity, and use that experience as the research object. In order to do that, to do more than simply return from the research with a one-off story to tell, it is necessary to declare in advance the intellectual framework you, the researcher, will use to try to make sense of the experience gained. Given such an explicit framework, you can then describe the research experience in the well-defined language of the framework. This makes it possible for anyone outside the work to 'recover' it, to see exactly what was done and how the conclusions were reached. This 'recoverability' requirement is obviously not as strong as the 'repeatability' criterion for scientific findings within natural science. But then, human situations are very much more complex than the phenomena studied in physics and chemistry labs! It is the declared framework and the recoverability criterion which clearly separate accounts of well-organized action research from novel writing – which, alas, too much published social research resembles.

In the action research which produced SSM the initial declared framework was the Systems Engineering approach developed by the Bell Telephone Company from their own case histories. Systems Engineering (SE) is a process of naming a 'system' (assumed to be some complex object which exists or could exist in the real world), defining its objectives, and then using an array of techniques developed in the 1950s and 1960s to 'engineer' the system to meet its objectives. This framework was rapidly found to be poverty-stricken when faced with the complexity of human situations. It was too thin, not rich enough to deal with fizzing social complexity.

The SE framework was modified (and enriched) in the light of and in direct response to real-life experiences. Eventually, we had in our hands an adequately rich framework, but it was far removed from the starting point in SE. It became known as Soft Systems Methodology. It then took some time for even its pioneers to realize just how radical the shift had been from SE to SSM. Having introduced the notion of 'worldview' – essential in dealing with human social complexity – we were thereafter thinking of systems models not as descriptions of something in the real world but simply as devices (based on worldview) to organize a debate about 'change to bring about improvement'. That was the key step in finding our way to SSM. This important shift in thinking is not abstruse, but it turns out to be very difficult for many people to grasp, simply because everyone is so used to the casual everyday-language use of the word 'system'. In ordinary talk we constantly refer to complex chunks of the everyday world as systems, even though they do not come close to meeting the requirements of that concept. We speak of 'the education system', 'health-care systems', 'the prison system', etc. using the word 'system' simply to indicate a chunk of reality which seems to be very complex but is, in some vague sense, a whole, something which might be better 'engineered'. Figure 1.7 gives a visual indication of the shift in thinking as SE was transformed into SSM.

At the starting point (SE) in Figure 1.7 (which ignores worldviews), 'systems' are names for things in the world which, given precise objectives, can be engineered to achieve them. At the end point (which accepts different world-

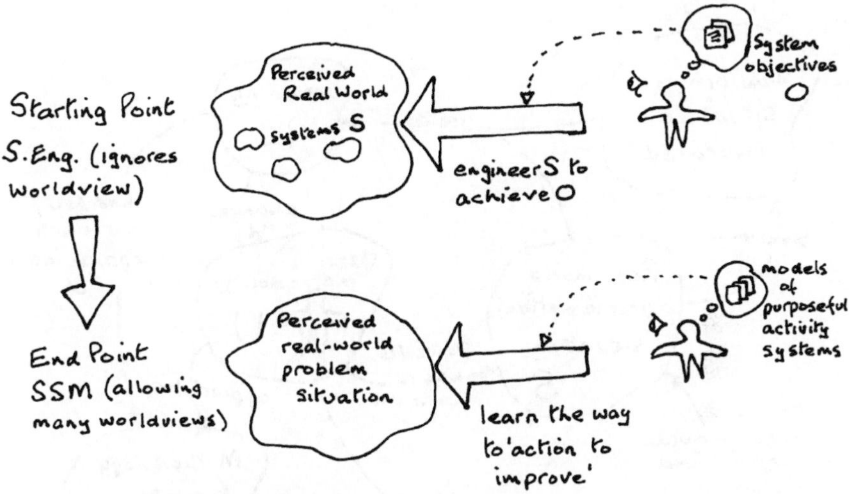

Figure 1.7 The shift in thinking entailed in developing SSM

views), 'systems' are devices used in a learning process to define desirable and feasible 'action to improve'.

Once the end point in Figure 1.7 was reached, and the SSM framework had been established, it was further developed, modified and honed in a few hundred new experiences. Out of this came a model which captures all of these developmental experiences. The model, known as the LUMAS model, is shown in Figure 1.8. (It is in fact a generic model for making sense of any real-world application of any *methodology*, remembering that that word covers a set of principles which need to be embodied in an application tailored to meet the unique features of a particular situation.)

LUMAS stands for Learning for a User by a Methodology-informed Approach to a Situation. In order to 'read' this model, start from the user (U) in the centre. He or she, perceiving a problem situation (S) and appreciating the methodology (M), tailors the latter to the former to produce the specific approach (A) to be used in this situation (S). This not only produces an improved situation but also yields learning (L). This will change the user,

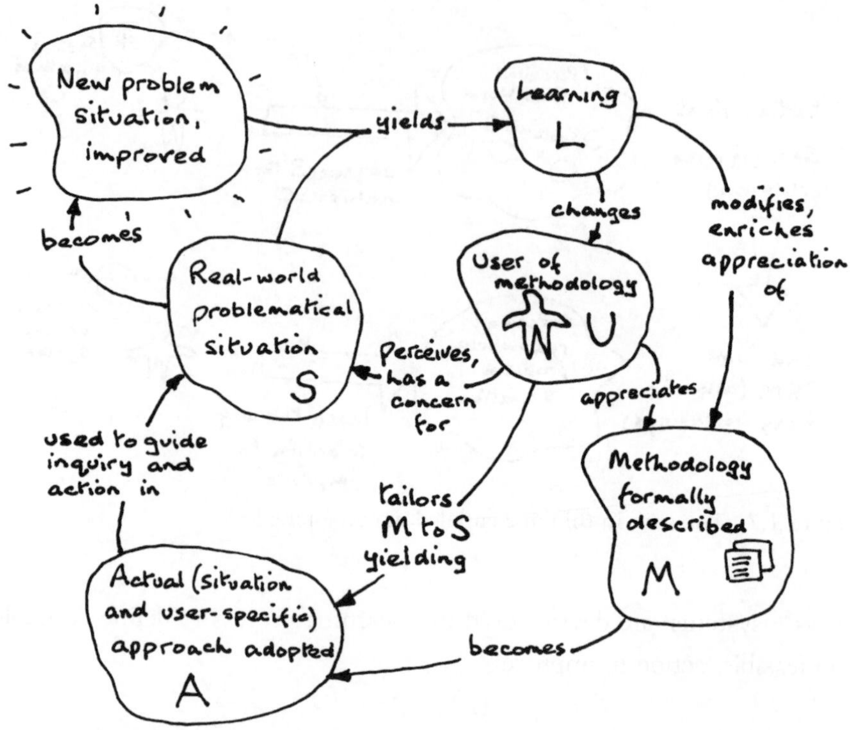

Figure 1.8 The LUMAS model – Learning for a User by a Methodologically-informed Approach to a Situation

who has gained this experience, and may also modify or enrich appreciation of the methodology. Every use of SSM can in principle be described in the language of this model. It is the gradually diminishing activity, over the years, of development occurring along the arrow which links L and M that makes it legitimate to describe SSM as mature.

How Does SSM Differ from Other Systems Approaches?

As described above, changes had to be made to Systems Engineering when it proved too blunt an instrument to deal with the complexity of human situations. Those changes explain SSM's difference from the other systems approaches developed in the 1950s and 1960s. SE is an archetypal example of what is now known as 'hard' systems thinking. Its belief is: the world contains interacting systems. They can be 'engineered' to achieve their objectives. This is the stance not only of SE; this thinking also underpins classic Operational Research, RAND Corporation 'systems analysis', the Viable System Model, early applications of System Dynamics and the original forms of computer systems analysis. None of these approaches pays attention to the existence of conflicting worldviews, something which characterizes all social interactions. In order to incorporate the concept of worldview into the approach being developed, it was necessary to abandon the idea that the world is a set of systems. In SSM the (social) world is taken to be very

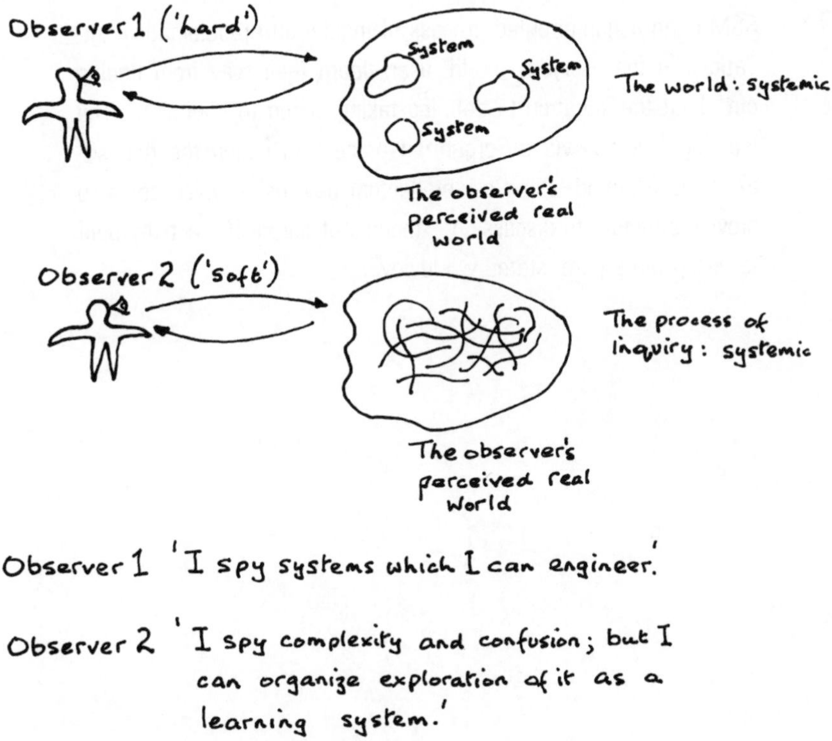

Figure 1.9 The 'hard' and 'soft' systems stances

complex, problematical, mysterious, characterized by clashes of worldview. It is continually being created and recreated by people thinking, talking and taking action. However, our coping with it, our process of inquiry into it, can itself be organized as a learning *system*. So the notion of systemicity ('systemness') appears in the process of inquiry into the world, rather than in the world itself. This shift created 'soft' as opposed to 'hard' systems thinking, the different stances adopted by the two being shown in Figure 1.9, itself another version of Figure 1.7.

This brings us to the end of a skeletal account of SSM as a whole. The next chapter expands on this, describing the techniques used in the cyclic process in detail. Meanwhile it seems worthwhile to try to summarize the broad account of SSM in a couple of sentences.

> SSM is an action-oriented process of inquiry into problematical situations in the everyday world; users learn their way from finding out about the situation to defining/taking action to improve it. The learning emerges via an organized process in which the real situation is explored, using as intellectual devices – which serve to provide structure to discussion – models of purposeful activity built to encapsulate pure, stated worldviews.

2
A Fleshed-out Account of SSM

Introduction

The previous chapter has answered the basic question about SSM, namely: What is it? And it has provided some context concerning its development, its application area and its crucial difference from the earlier systems approaches from the 1950s and 1960s. In this chapter the focus is on 'how' rather than 'what': How exactly does the user move through the learning cycle of SSM, shown in Figure 1.5, in order to define useful change? Which techniques for finding out, modelling and using models to question the real situation have shown themselves robust enough to survive in many different circumstances, so that they have become part of the classic approach?

The account here will follow the four basic activities of the broad-brush account (finding out, modelling, using the models to structure debate, and defining/taking action), with the usual reminder that activity in any project using SSM will reflect the kind of pattern shown in Figure 1.6 rather than a stately linear progress.

The SSM Learning Cycle: Finding Out

Four ways of finding out about a problematical situation have survived many tests and become a normal part of using SSM. In the language of SSM they are known as 'making Rich Pictures' and carrying out three kinds of inquiry, known as 'Analyses One, Two and Three'. These focus, respectively, on the intervention itself, a social analysis (What kind of 'culture' is this?) and a political analysis (What is the disposition of power here?). They will be described in turn.

(Readers anxious to reach the stories of SSM use might turn to the first few case histories described in Part Two, but all the accounts there use the terms and language carefully defined here, so a little patience might well be worthwhile!)

Making Rich Pictures

Entering a real situation in order first to understand it and then to begin to change it in the direction of 'improvement' calls for a particular frame of mind in the user of SSM. On the one hand the enquirer needs to be sponge-like, soaking up as much as possible of what the situation presents to someone who may be initially an outsider. On the other hand, although holding back from imposing a favoured pattern on the first impressions, the enquirer needs to have in mind a range of 'prompts' which will ensure that a wide range of aspects are looked at. Initially two dense and cogent questions were used as a prompt:

- What resources are deployed in what operational processes under what planning procedures within what structures, in what environments and wider systems, by whom?
- How is resource deployment monitored and controlled?

Certainly, if you can answer these questions you know quite a lot about the situation addressed. But these questions did not survive as a formal part of SSM. (The problem with them is that when they were formulated, in the early days of SSM development, the thinking of the pioneers had not sufficiently divorced itself from thinking of the world as a set of systems.

The questions imply intervention in some real-world system – hence the references to 'wider systems' and to monitoring and control – rather than the intervention being addressed to *a situation*.) The questions would no doubt have been changed eventually as the true nature of SSM was realized. However, what happened instead was that the questions were dropped because the phrase 'rich picture' quickly moved from being a metaphor to being a literal description of an account of the situation *as a picture*.

The rationale behind this was as follows. The complexity of human situations is always one of multiple interacting relationships. A picture is a good way to show relationships; in fact it is a much better medium for that purpose than linear prose. Hence as knowledge of a situation was assembled – by talking to people, by conducting more formal interviews, by attending meetings, by reading documents, etc. – it became normal to begin to draw simple pictures of the situation. These became richer as inquiry proceeded, and so such pictures are never finished in any ultimate sense. But they were found invaluable for expressing crucial relationships in the situation and, most importantly, for providing something which could be tabled as a basis for discussion. Users would say: 'This is how we are seeing your situation. Could we talk you through it so that you can comment on it and draw attention to anything you see as errors or omissions?'

In making a Rich Picture the aim is to capture, informally, the main entities, structures and viewpoints in the situation, the processes going on, the current recognized issues and any potential ones.

Here is a real-world problematical situation described in a paragraph of prose:

> The newly appointed headteacher of an 11s-to-18s school, which has overspent its budget in the last year or two, finds herself, in her first term, facing an issue concerning the provision of school meals. Currently these are provided by the county education authority

through their catering services company, the contract being renewed annually. A member of that company who is leaving to set up her own catering company urges the headteacher to make a contract with her instead of the county, suggesting the school could save money on this. Some staff members agree with this, others want to stick with the status quo. Some parents, alerted by a national debate about school meals, want more nutritious meals as long as they don't cost more. Pupils say: 'We like burgers and chips.' The school governors are discussing this issue; the Chairman, himself MD of a catering company, is urging the headteacher to be entrepreneurial and to take on responsibility for the provision of school meals, believing this could be profitable for the school.

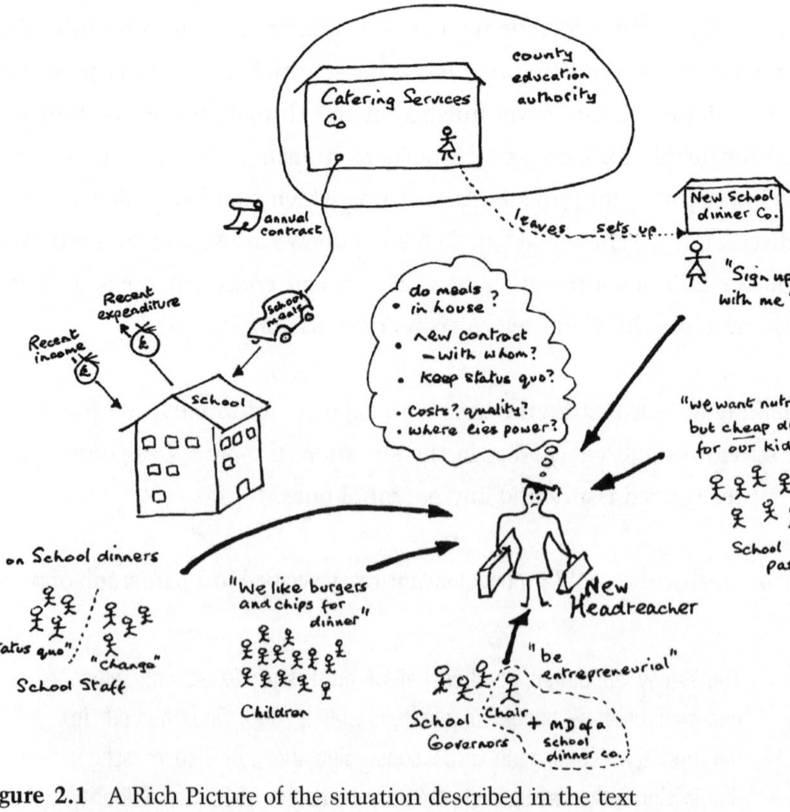

Figure 2.1 A Rich Picture of the situation described in the text

Figure 2.1 represents this situation in a Rich Picture. Our point is that this picture is a more useful piece of paper than the prose account. It could lead to a better-than-usual level of discussion because not only can it be taken in as a whole, but also it displays the multiple *relationships* which the headteacher has to manage, not just immediately, but through time. That is the power of such pictures, though we have to remember that however rich they are they could be richer, and that such pictures record a snapshot of a situation which will itself not remain static for very long. Wise practitioners continually produce such pictures as an aid to thinking. They become a normal way of capturing impressions and insights.

Carrying Out Analysis One (the Intervention Itself)

Whenever SSM is used to try and improve a problematical situation three elements – the methodology, the use of the methodology by a practitioner and the situation – are brought together in a particular relationship, namely that shown in Figure 2.2. The practitioner will adapt the principles and

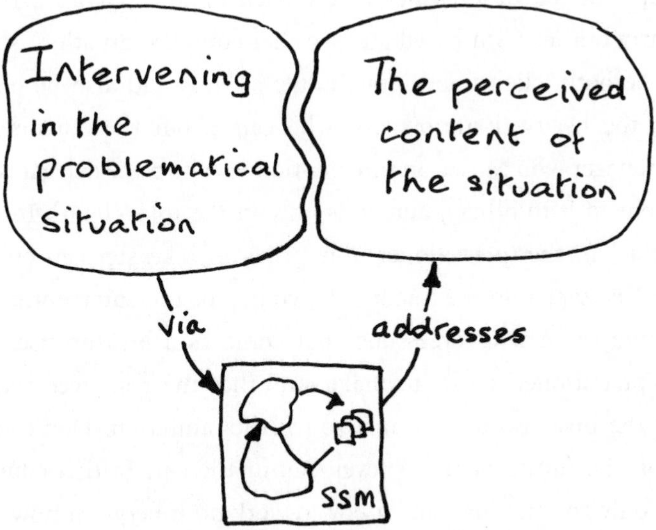

Figure 2.2 The three elements in any SSM investigation

techniques of the methodology to organize the task of addressing and intervening in the situation, aiming at taking action to improve it. In developing SSM, this process was organized in a sequence of real situations, and it was quickly found useful to think about Figure 2.2 in a particular way. Three key roles were always present:

1. There was some person (or group of persons) who had *caused the intervention to happen*, someone without whom there would not be an investigation at all – this was the role 'client'.
2. There was some person (or group of persons) who were *conducting the investigation* – this was the role 'practitioner'.
3. Most importantly, whoever was in the practitioner role could choose, and list, a number of people who could be regarded as being *concerned about or affected by the situation and the outcome* of the effort to improve it – this was the role 'owner of the issue(s) addressed'.

It is important to see why these are named as 'roles' rather than particular people. It is because one person (or group) might be in more than one role. For example, if the headteacher in the Rich Picture (Figure 2.1) were to herself carry out an SSM-based study of her complex situation, she would not only be both 'client' and 'practitioner', she would also be one of the people in the list of 'issue owners' who care about the outcome. Sometimes a manager who causes an intervention to take place delegates detailed involvement in it to others, and so is only in the role 'client'. In this case the person(s) in the 'practitioner' role needs to take steps to ensure that the 'client' is kept informed about the course of the intervention so that the outcome when it emerges does not come as a big surprise. In every case the 'practitioner' needs to make sure that the resources available to carry out the investigation are in line with its ambition. Don't undertake a study of 'the future of the A-level examination in British education' if you have only got one man and a boy to work on it between now and next Thursday.

SSM's 'Analysis One', then, consists of thinking about the situation displayed in Figure 2.2 in the way shown in Figure 2.3, asking: Who are in the roles 'client' and 'practitioner'? and Who could usefully be included in the list of 'issue owner'?

Much learning came out of the simple thinking which led to this 'Analysis One'. For example, it was always useful to think about the client's aspirations for the intervention. They should always be taken seriously but should not be the sole focus of the work done. Thus, the person(s) in the 'client' role should be in the list of possible 'issue owners' but should very definitely not be the only one in the list. In this connection it was interesting to hear a

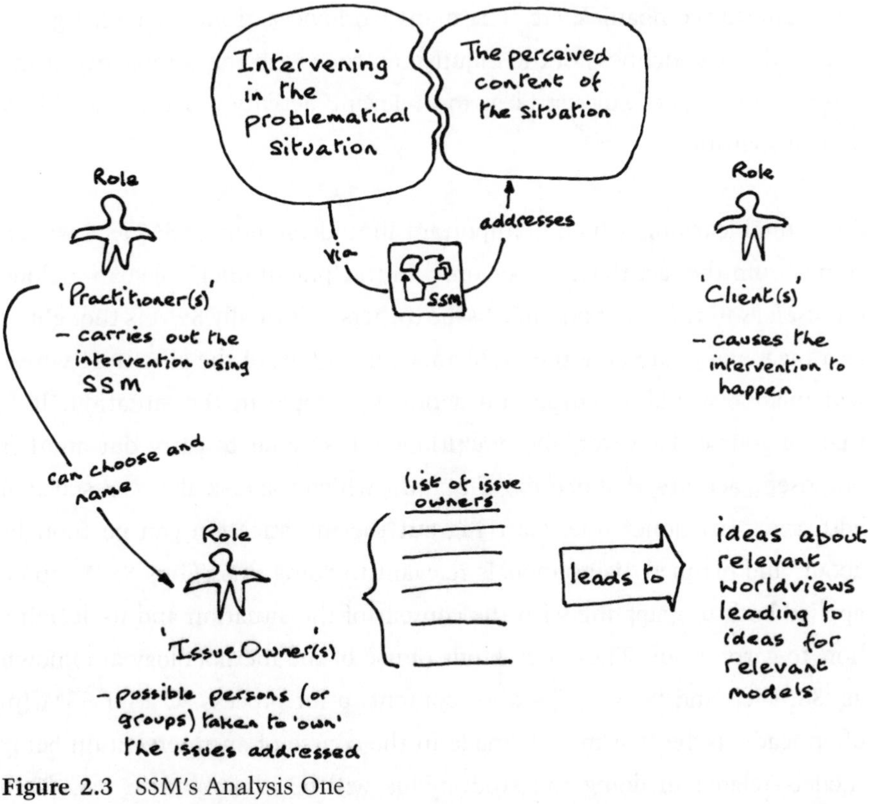

Figure 2.3 SSM's Analysis One

senior manager from the RAND Corporation declare, some years ago, 'The RAND analyst places his or her expertise at the disposal of a real-world decision-taker who has to be a legitimate holder of power.' In the language of Figure 2.3 this was to declare that for RAND the client *is* the issue owner, full stop. This cuts off all the richness which comes from the practitioner compiling a list of persons or groups who *could be taken to be* issue owners; for it is that list which introduces multiple worldviews. They in turn open up the chance of a richness of learning at a deep level for all involved in the intervention, leading, perhaps, to major change. The RAND manager's statement would define the practitioner as only a servant to the legitimately powerful. In the situation shown in Figure 2.1, for example, 'issue owners' might include: the headteacher; the school governors, staff and pupils; parents; the county education authority and their catering services company; other catering companies, etc. The many worldviews from such a list give a chance that the richness of the inquiry can cope with the complexity of the real situation. They suggest ideas for 'relevant' activity models, ones likely to be insightful.

Some final learning, which is important in understanding SSM as a whole, comes from the fact that the person(s) in the 'practitioner' role can include *themselves* in the list of possible 'issue owners'. Normally SSM is thought of as a means of addressing the problematical content of the situation, which will include would-be purposeful action by people in the situation. It *is* that, of course. However, the practitioner(s) is about to carry out another purposeful activity, that of *doing the study*, which is a task always associated with the practitioner role. Carrying out the investigation can be thought about, and planned, using models relevant to doing this. Thus SSM can be applied both to grappling with the content of the situation and to deciding how to carry it out. These two kinds of use of the methodology are known as 'SSM (c)' and 'SSM (p)' – c for content, p for process. Use of SSM (p) often leads to the first models made in the course of an intervention being models related to doing the study. This will be illustrated in Part Two

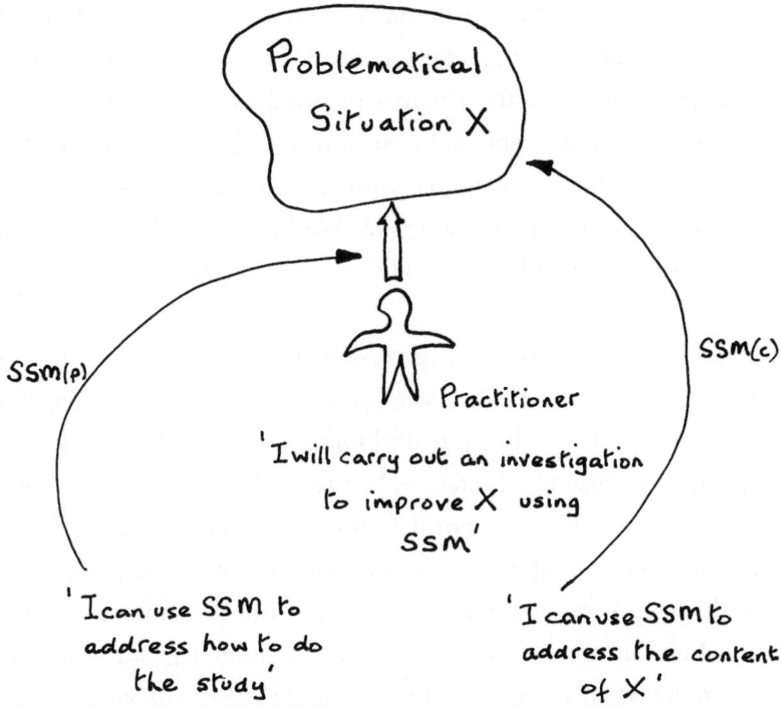

Figure 2.4 SSM(p) concerned with the process of using SSM to do the study and SSM(c) concerned with the problematical content

(Chapter 3, Case 1, Figure 3.1). Meanwhile Figure 2.4 illustrates these two ways of using SSM.

Carrying Out Analysis Two (Social)

It might seem obvious that if you are going to intervene in, and change, a human situation, you ought to have a clear idea about what it is you are intervening in. You should have some sense of what you take 'social reality' to be. However, this is not too obvious! The Management Science field, for example, tries to get by through concentrating almost entirely on the *logic* of situations, even though the motivators of much human action lie outside logic, in cultural norms or emotions. So, if we are to be effective in social situations, we have to take 'culture' seriously and decide what we mean by it. This is especially important for SSM as an action-oriented approach. If we are to learn our way to practical action which will improve a situation under investigation, then the changes involved in 'improvement' have to be not only arguably desirable but also *culturally feasible*. They need to be possible for these particular people, with their particular history and their particular ways of looking at the world. We have to understand the local 'culture', at a level beyond that of individual worldviews.

This might be straightforward if there were an agreed definition of exactly what we mean by 'culture'. However, there is no agreed definition, though the concept is much discussed by anthropologists, sociologists and people writing in the management literature. By the 1950s, a survey (by Kluckhohn and Kroeber) found 300 different definitions, and no agreement has been reached since then! In spite of that, everyone has a general, diffuse sense of what the word means. If you say 'This is a "can-do" culture', or 'This is a buttoned-up culture', or assert that 'The Civil Service is a punishment-avoiding, rather than a reward-seeking culture' then it will be accepted that you have said something meaningful. To anyone familiar with the society in question, those statements will have conveyed some sense of the 'feel', or 'flavour', of the situation: its social texture. In order to pin down such feelings more firmly, in a way which makes practical sense, SSM makes use of a particular model. This is a model which does not claim the status of rounded theory, but it has proved itself useful in situations from small firms dominated by individuals to large corporations which develop and (partially) impose their own norms.

The model is at the same time simple (you can keep it in your head) but also subtle. It consists of only three elements – roles, norms, values – but the subtlety comes from the fact that none of these elements is static. Each, over time, continually helps to create and modify the other two elements, as shown in Figure 2.5.

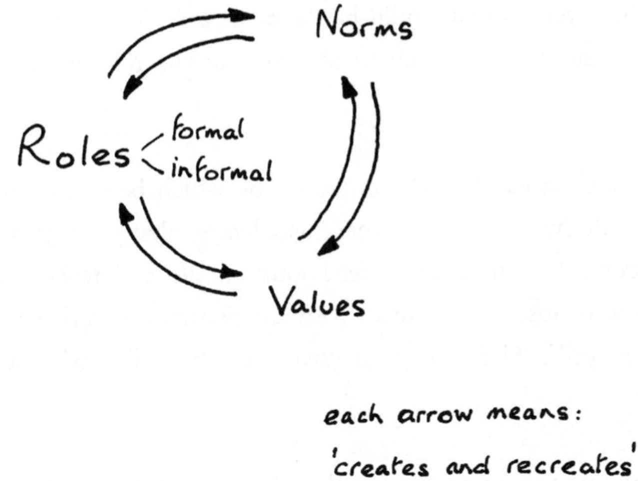

Figure 2.5 SSM's model for getting a sense of the social texture of a human situation

Together the three elements help to create the social texture of a human situation, something which will both endure *and* change over time. Consider the three elements in turn.

Roles are social positions which mark differences between members of a group or organization. They may be formally recognized, as when a large organization has, say, a chief executive, directors, department heads, section heads and members of sections. But in any local culture informal roles also develop. Individuals may develop a reputation as 'a boat-rocker', or 'a licensed jester' – someone who can get away with saying things others would suppress. The informal roles which are recognized in a given culture tell you a lot about it.

Norms are the expected behaviours associated with, and helping to define, a role. Suppose you told a friend you were going to meet 'the vice-chancellor of a UK university' next day. If you returned from the meeting and said that the VC sat picking her teeth, with her feet on the table, and was very foul-mouthed, your friend would be flabbergasted. Such behaviour is way outside the expected behaviour of someone in the role of VC in British society.

Values are the standards – the criteria – by which behaviour-in-role gets judged. In all human groups there is always plenty of gossip related to this. People love to discuss behaviour in role and reach judgements which praise or disparage: 'He's a very efficient town clerk who services committees well'; 'She's an ineffective vice-chancellor who won't take decisions.'

It is obvious from these definitions that the three elements – roles, norms, values – are closely related to each other, dynamically, and that they change over time as the world moves on. Anyone who has ever been promoted within an organization will know that occupying the new role changes them, as they adopt a new perspective appropriate to the role. Equally, how they enact the new role will have its effect, in future, on the local norm – the behaviour which people expect from whoever fills that role. The elements also change over time at a macro level. For example, when the authors were growing up in British society the worst role for a young woman to find herself in was to be an unmarried mother. At that time, society judged harshly the behaviour which led to this. Not any more; the social stigma attached to the role has disappeared in the UK over the last 50 years.

So how exactly is the model of linked roles, norms and values in Figure 2.5 used in SSM? At the start of an intervention open a file marked 'Analysis Two'. Then, every time you interact with the situation – talking to people informally, reading a document, sitting in a meeting, conducting an interview, having a drink in the pub after work – ask yourself afterwards whether that taught you anything about the roles, norms and values which are taken seriously here and characterize this particular group. Record the finding in the 'Analysis Two' file. Carry on doing this throughout the engagement, and

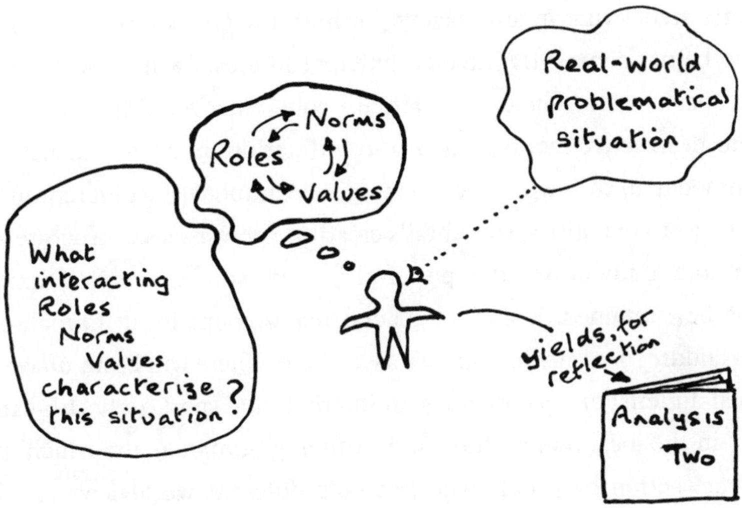

Figure 2.6 SSM's Analysis Two

put a date on every entry so that later on you can recover the progress of your learning, and reflect upon it. Figure 2.6 summarizes Analysis Two.

Carrying Out Analysis Three (Political)

The experienced reader will have noticed that so far in this discussion of 'Finding Out' about a problematical situation we have made no mention of the *politics* of a situation, which are always powerful in deciding what does or does not get done. That is the focus of Analysis Three: to find out the disposition of power in a situation and the processes for containing it. That is always a powerful element in determining what is 'culturally feasible', politics being a part of culture not addressed directly in the examination of roles, norms and values of Analysis Two.

The 'political science' literature contains many models – usually fairly complex ones – which set out to express the nature of politics. The model used in SSM, in Analysis Three, does not come from that literature but from some basic ideas found in the work of the founding father of the field: Aristotle.

Aristotle argues that in any society (for him, the Greek city-state) in which human beings constantly interact, different interests will be being pursued. If the society as a whole is to remain coherent over time, not breaking up into destructive factions, then those differing interests will have to be accommodated; they will never go away. Accommodating different interests is the concern of politics; this entails creating a power-based structure within which potentially destructive power-play in pursuit of interests can nevertheless be contained. This is a general requirement in all human groups which endure, not only in societies as a whole. There will be an unavoidable political dimension in companies, in international sport, in health-care provision, in the local tennis club – in fact in any human affairs which involve deliberate action by people who can hold different worldviews and hence pursue different interests.

Analysis Three in SSM asks: How is power expressed in this situation? This is tackled through the metaphor of a 'commodity' which embodies power. What are the 'commodities' which signal that power is possessed in this situation? Then: What are the processes, by which these commodities are obtained, used, protected, defended, passed on, relinquished, etc? Figure 2.7 summarizes Analysis Three. The commodities which indicate power in human groups are, of course, many and various. There is a link here to Analysis Two, since occupying a particular role embodies power: the chief constable has more power than a detective sergeant, by virtue of his role. Other common commodities of power include, for example: personal charisma; membership of various committees in organizations; having regular access to powerful role-holders; in knowledge-based settings, having intellectual authority and reputation; having authority to prepare the minutes of meetings – a chore, perhaps, but it gives you some power! Many commodities of power derive from information. Having access to important information, or being able to prevent others from having access to certain information, is a much-used commodity of power in most organizations.

A dramatic example of an unusual commodity of power in a specific SSM project was revealed when two managers in a consultancy company were being interviewed as a pair. They began to disagree with each other and, in a deliberate bit of power-play, one of them suddenly said: 'You say that, but you're NKT; I'm KT.' This local private language within this company referred to those partners who 'knew Tom' and those, more recent joiners, who 'never knew Tom', Tom being the charismatic founder of the company, now deceased. This taught those facilitating this use of SSM that there was an unstated but very real hierarchy here. The KTs, Tom's original disciples, were much more influential than the come-lately NKTs. This indicated that the only changes likely to be culturally feasible in this situation would be those supported by the KTs, whose power stemmed from their association with the charismatic Tom. This is an interesting example of a commodity of power which would gradually fade over time. And this itself reminds us that, as with Analysis Two, Analysis Three deals with elements which are continually being redefined as life moves on.

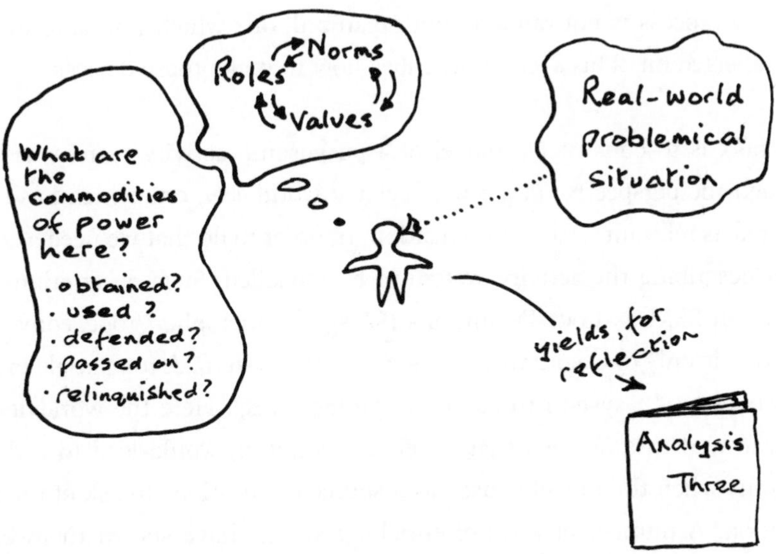

Figure 2.7 SSM's Analysis Three

The way of doing this analysis echoes that of Analysis Two: open a file and record in it – with a date – any learning gained about power and the processes through which it is exercised. Do this, and reflect upon it, over the whole course of an investigation.

The SSM Learning Cycle: Making Purposeful Activity Models

As explained in Chapter 1, in order to ensure that learning can be captured, SSM users create an *organized* process of enquiry and learning. They do this by making models of purposeful activity and using them as a basis for asking questions of the real-world situation. This kind of model is used because every human situation reveals people trying to act purposefully. Since each model is built according to a declared single worldview (e.g. 'the Olympic Games from the perspective of the host city') such models could never be descriptions of the real world. They model *one way* of looking at complex reality. They exist only as devices whose job is to make sure the learning process is not random, but organized, one which can be recovered and reflected on. This section describes how to make these devices.

The task is to construct a model of a purposeful 'activity system' viewed through the perspective of a pure, declared worldview, one which has been fingered as relevant to this investigation. In order to do that we need a statement describing the activity system to be modelled. Such descriptions are known in SSM as Root Definitions (RDs), the metaphor 'root' conveying that this is only one, core way of describing the system. A too-simple example would be: 'A system to paint the garden fence'. Here the worldview is unclear, and it is obvious that a richer description would lead to a richer outcome when the model is used as a source of questions to ask of the real situation. A number of ways of enriching an RD have shown themselves to be useful. For example, we could more richly express the RD above as:

'A householder-owned and staffed system to paint the garden fence, by hand-painting, in keeping with the overall decoration scheme of the property in order to enhance the appearance of the property'. This makes clear that the model takes a householder's worldview as given, and that that particular householder believes in DIY activity to improve it. In addition it not only describes *what* the system does (paint the fence); it also says *how* (by hand-painting) and *why* (to enhance the appearance of the property). (Also the worldview assumes a link between painting and improving appearance.) Clearly this would lead to a richer questioning of the real situation to which this purposeful activity was thought to be relevant as a device to structure the questioning.

The whole set of guidelines of this kind – there to help the modelling process – will now be described. They are set out in Figure 2.8; the five numbered elements in the figure will be described in turn.

1. The formula followed in enriching the fence-painting RD above is always helpful, and can apply to every RD ever written. It is known in SSM as 'the PQR formula': do P, by Q, in order to help achieve R, where PQR answer the questions: What? How? and Why? PQR provides a useful shape for any and every RD. Remember, though, in using PQR, that if the formula is complete, with all three elements defined, then the transforming process is captured in Q, the declared 'how'. In the simple example above the Q is 'hand-painting' (not simply 'painting'). Also, though it is not an issue in this example, the model builder has to be able to defend Q as a plausible 'how' for the 'what' defined by P. If you were to write 'define health-care needs' as P and then define Q *only* as 'by asking patients for their views' this would not be easily defensible.

2. The PQR formula allows you to write out the RD as a statement. This always describes the purposeful activity being modelled as a transformation process, one in which some entity (in the example an 'unpainted fence') is transformed into a different state (here, a 'painted fence'). Any purposeful

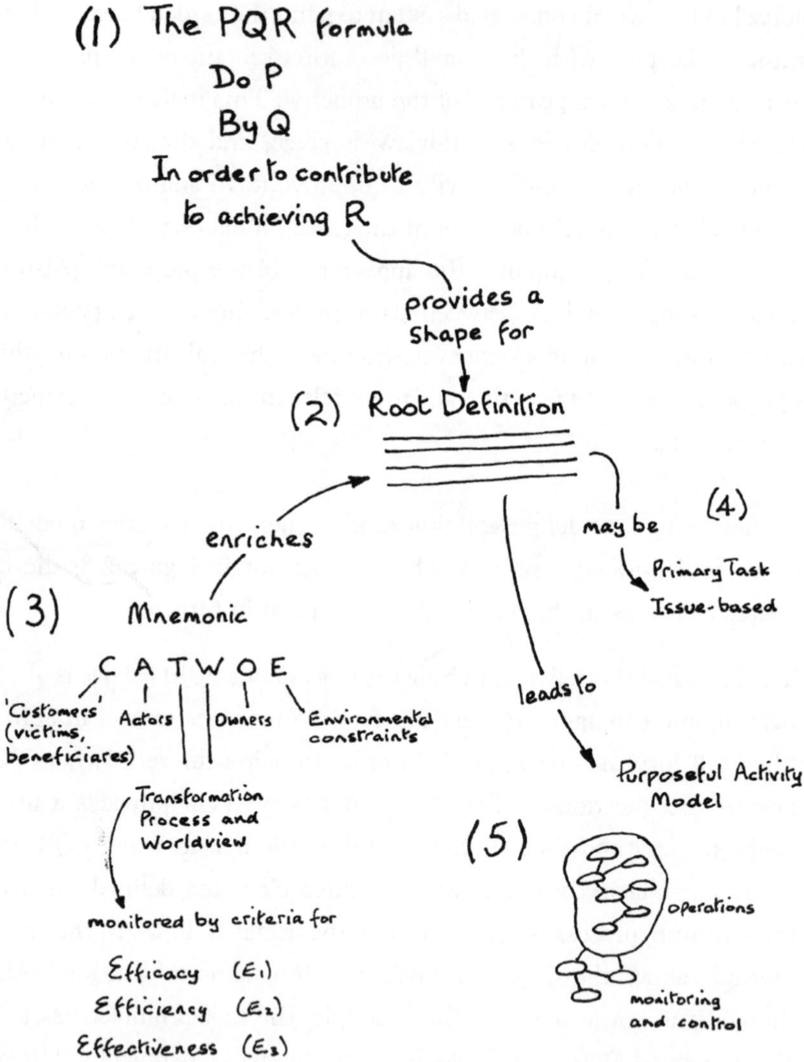

Figure 2.8 Guidelines which help with building models of purposeful activity

activity you can think of can be expressed in this way, which is useful because it makes model building a straightforward process. For complex activities the entity being transformed will probably be best expressed in an abstract way, for example: 'the health-care needs of Coketown citizens' transformed into 'the health-care needs of Coketown citizens met'. But the idea of purposeful activity as a transformation always holds, whether the transformation is concrete or abstract. Putting together the activities needed to describe the transforming process (i.e. 'building the model') can begin when an RD is complete, but before moving on to this, elements 3 and 4 in Figure 2.8 should be considered. They further enrich the modelling and improve it as a source of questions to ask in the real situation.

3. When the idea of working with RDs as a source of models was being developed, a further enrichment of the thinking came from having, as a reference, a completely general model of any purposeful activity. (This was a way of declaring exactly what we meant by 'purposeful activity'.) The general model is shown in Figure 2.9. It contains elements which can usefully be thought about for any purposeful (transforming) activity.

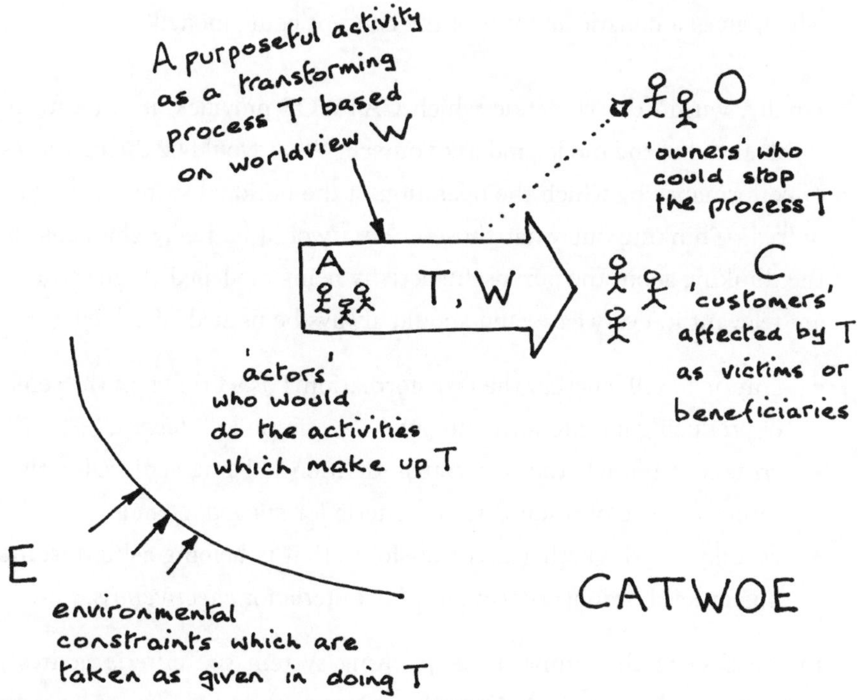

Figure 2.9 A generic model of any purposeful activity, which yields the mnemonic CATWOE

The model provides the mnemonic CATWOE, defined as in Figure 2.9. The concept here is that purposeful activity, defined by a transformation process and a worldview (a T and a W):

- will require people (A) to do the activities which make up T;
- will affect people (C) outside itself who are its beneficiaries or victims (C for 'Customers');
- will take as given various constraints from the environment outside itself (E) (such as a body of law, or a finite budget);
- could be stopped or changed by some person or persons (O) who can be regarded as 'owning' it.

Many people find it useful, when model building, to start the process by defining first T and W, then the other CATWOE elements. Experience suggests, though, that it is still useful to write out the RD as a statement which gives a holistic account of the concept being modelled.

Finally, within the guideline which CATWOE provides, it is useful to think ahead to the model and ask yourself: What would be the measures of performance by which the operation of the notional system would be judged? Thinking out what those criteria would be really sharpens up the thinking about the purposeful activity being modelled. Three criteria are relevant in every case, and should always be named. We need:

- criteria to tell whether the transformation T is working, in the sense of producing its intended outcome, i.e. criteria for *efficacy*;
- criteria to tell whether the transformation is being achieved with a minimum use of resources, i.e. criteria for *efficiency*; and
- criteria to tell whether this transformation is helping achieve some higher-level or longer-term aim, i.e. criteria for *effectiveness*.

In the case of the simple fence-painting system the criteria address, respectively, the questions: Does this count as 'a painted fence' (human judgement would decide)? Is the painting being done with minimum use of the resources of materials and time (these might be expressed as costs)? and Does the painted fence enhance the appearance of the property (again human judgement would decide)? These three criteria are always independent of each other. Thus, for example, the purposeful act of taking a drug to relieve your headache might be efficacious if the headache goes. But it could be inefficient if the drug cost too much or was very slow-acting. And it could also be ineffective, medically, if treating the symptom of the headache was unwise because the headache actually signalled a more serious complaint.

These 'three Es' will always be relevant in building any model, but in particular circumstances other criteria might also apply, such as *elegance* (Is this a beautiful transformation?) or *ethicality* (Is this a morally correct transformation?). The judgement is yours as to what criteria are needed.

4. The final consideration in Figure 2.8 when formulating RDs prior to model building concerns RDs as a whole. Are they 'Primary Task' or 'Issue-based' definitions? This useful distinction (though it does not affect model building technique) arose through experience, like most developments in SSM. In the early days, when the legacy of Systems Engineering hung heavy over the new approach, the models built were always of purposeful activity of a kind that was present in the real world in the form of departments, divisions, sections, etc.; that is to say it was institutionalized. Thus, if working in a company with functional sections – production, marketing, research and development, etc. – we would in the early days of developing SSM make models only of a production system, a marketing system, an R&D system, etc. In these cases the boundary of the models we built would coincide with internal organizational boundaries. This is not 'wrong', but it puts limitations on the thinking of the team carrying out the investigation, which may go unnoticed. Every organization has to carry out many, many purposeful activities as it goes about its business. Only a few of these can be

captured in the organization structure as departments, etc. These organizational boundaries are, in the last analysis, arbitrary, and could be changed.

Experience quickly showed that to stimulate the thinking of everyone involved in the investigation it was useful to make models of purposeful activity whose boundaries *cut across organizational boundaries*. These are 'Issue-based' models from 'Issue-based' RDs, models whose boundaries do not coincide with organizational boundaries. When such models are used to ask questions in the situation, interest and attention are always increased. This brings in broader considerations than is the case with a model which accepts organizational boundaries as a given. This is because the questions about what departments, sections, etc. should exist, and what their boundaries should be are always bound up in the power-play going on in organizations. That catches everyone's attention!

As a generalization we can suggest one choice of Issue-based RD which is always worth considering. In virtually all organized human groups there will always be contentious issues concerned with allocating resources. This is something which affects all members, leads to wide discussion, and is not usually assigned as an activity to a particular sub-group. An issue-based model based on transforming unallocated into allocated resources will be worth considering as a stimulant in most investigations. The general rule is: never work exclusively with either Primary Task (PT) or Issue-based (IB) RDs. Most investigations will best feature a mixture of both types.

5. Earlier in this chapter, in section 2 above, model building was described as 'putting together the activities needed to describe the transforming process', in other words defining and linking the activities needed to achieve the transforming process. Given the guidelines provided by PQR, an RD, CATWOE, the 3Es and PT/IB, this task should not be a difficult

one. The only skill called for is logical thinking. The most common error – even among logical thinkers – is to take your eye off the root definition and start modelling some real-world version of the purposeful activity being modelled. In work in a medium-sized manufacturing company, concerned with various issues regarding product distribution, it was easier for the SSM practitioners to build relevant models than it was for the distribution manager. He kept slipping into modelling the current ways of working in his department rather than the concepts in RDs. If you do this, of course, you find yourself not questioning current practice but comparing X with X – not very profitable!

People find their own way of making the selected relevant models, but a logical sequence to follow, or to refer to if in difficulty, is as follows:

(1) Assemble the guidelines: PQR, CATWOE, the RD, etc.
(2) Write down three groups of activities – those which concern the thing which gets transformed (the 'unpainted fence', or the 'health needs of the citizens of Coketown', in the examples above); those activities which *do* the transforming; and any activities concerned with dealing with the transformed entity (e.g. judging if it improves the appearance of the property, in the fence-painting example); this will give you a cluster of activities.
(3) Connect the activities by arrows which indicate the dependency of one activity upon another; for example, you can't *use* a raw material to make something before you've *obtained* it, so an arrow goes from an 'obtain' activity to the 'use' activity. In Figure 1.3 activity 7 (paint the fence) depends upon both activities 4, 5 and 6, since you can't paint the fence until you've obtained both brush and paint and prepared the fence.
(4) Add the three monitoring and control activities, which always have the structure shown in Figures 1.2 and 1.3.
(5) Check the model against the guidelines. Ask yourself: Does every phrase in the RD lead to something in the model? And: Can every activity in

the model be linked back to something in the RD or CATWOE, etc.? If the answer to both questions is 'Yes', then you have a defensible model. Note that the word used here is 'defensible' rather than 'correct'. This is because everyday words have different connotations for different people. Competent SSM practitioners working from the same RD might well produce somewhat different models; this is because they are interpreting the words in the RD, etc. somewhat differently. The important thing is that you can defend your model as representing what is in your RD, PQR, CATWOE, etc.

Figure 2.10 summarizes the model building process.

Finally, on model building, there is one more guideline worth taking seriously. *Aim* to capture the activity in the operational part of the model in 'the magical number 7 ± 2' activities (but do break the 'rule' if necessary). This famous phrase comes from a celebrated paper in cognitive psychology. George Miller, based on laboratory work, suggests that the human brain may have the capacity to cope with around seven concepts simultaneously. Whether or not this is true it is certainly the case that a set of 7 ± 2 activities can be thought about holistically. If the number seems low, this is not a problem. Any activity in a model can itself, at a more detailed level, become the source of an RD and a model. Thus, in Figure 1.3, activity 6 (obtain paint) could itself be expanded into a model which set out the connected, more-detailed activities which together combine to constitute 'obtain paint' – activities concerned with checking out suppliers, their prices, selecting one, etc. If this model were built, its activities would be numbered 6.1, 6.2, 6.3, etc. since they all derive from activity 6 in the parent model. In this way coherence is maintained no matter how many levels it may be necessary to go to in a particular investigation. In the authors' experience of more than a hundred studies it has never been necessary to expand beyond two levels below that of the parent model, and even then expanding only a few activities at the lower levels.

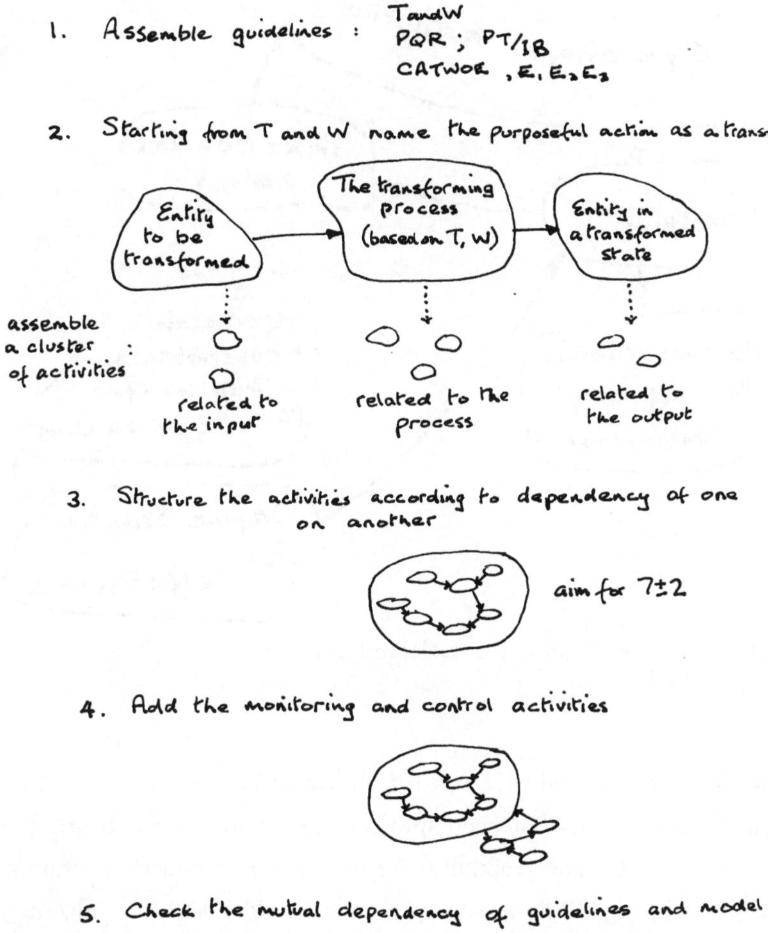

1. Assemble guidelines :
TandW
PQR ; PT/IB
CATWOE , E1 E2 E3
2. Starting from T and W name the purposeful action as a transformation:
Entity to be transformed
The transforming process (based on T, W)
Entity in a transformed state
assemble a cluster of activities :
related to the input
related to the process
related to the output
3. Structure the activities according to dependency of one on another
aim for 7±2
4. Add the monitoring and control activities
5. Check the mutual dependency of guidelines and model

Figure 2.10 A logical process for building SSM's activity models

The first model presented here, to illustrate the idea of purposeful activity models, was that in Figure 1.3. This was presented without a Root Definition, but now that this has been defined (above) we can present part of the model in a more developed form. This is done in Figure 2.11 which makes one particular change. It would have been possible to include in the 'operations' part of the model an activity such as 'ascertain the judgement about the enhanced appearance of the property'. Another way of bringing in the R of PQR (the higher-level, or longer-term aim of the transforming process,

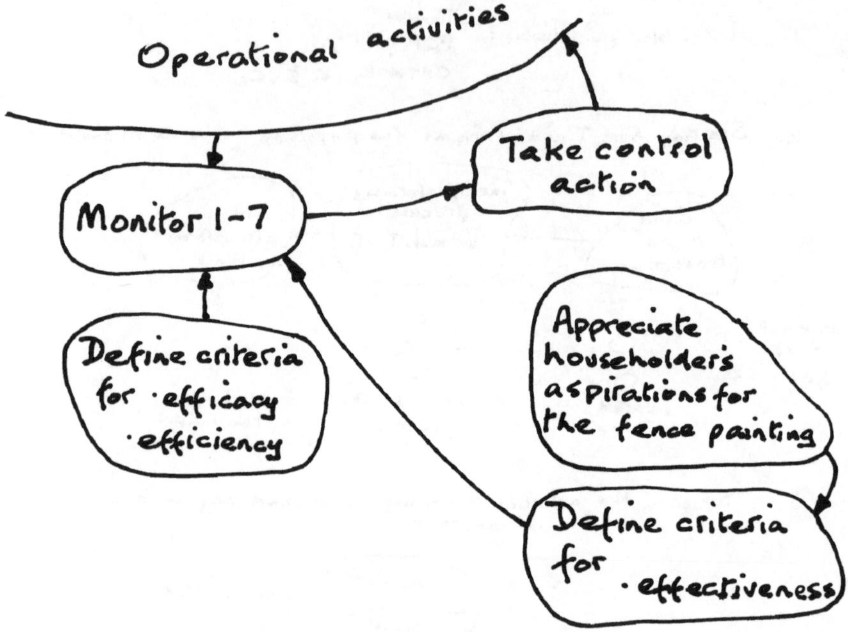

Figure 2.11 A variant of part of the model in Figure 1.3

judged by the criteria for effectiveness) is shown in Figure 2.11. The monitoring and control activity has been split into two, with the monitoring for effectiveness having the added activity: 'Appreciate householder's aspirations for the fence painting.' This leaves open who would make the judgement about the hoped-for enhancement of the appearance of the property – the householder? his or her partner? the neighbours? a prospective purchaser? This is probably, in this instance, the most elegant way of bringing all the elements in the guidelines into the model.

Appendix C carries an example of model building which starts from a Root Definition and 'talks through' the whole process for that definition. In general, the best way to learn about activity modelling is to have a look at examples and then have a go. And do remember that even rough-and-ready models can be helpful in real situations.

The SSM Learning Cycle: Using Models to Structure Discussion about the Situation and its Improvement

When we enter a problematical situation and start drawing rich pictures and carrying out preliminary versions of Analyses One, Two and Three, we begin to build up what can become a rich appreciation of the situation. This appreciation – helped especially by the list of possible 'issue owners' from Analysis One – enables us to begin to name some models which might be helpful in deepening our understanding of the situation and beginning to learn our way to taking 'action to improve'. Having built a hopefully relevant model or two, we are then ready to begin the structured discussion about the situation, and how it could be changed, which will eventually lead to action being taken. The models are the devices which enable that discussion to be a structured rather than a random one.

In everyday situations, typical discussions among professionals are characterized by a remarkable lack of clarity. In a typical 'management' discussion in an organization, unless there is a chairperson of near-genius, different voices will be addressing different issues; different levels, from the short-term tactical to the long-term strategic, will be being addressed; different speakers will assume different timescales. The resulting confusion will then provide splendid cover for personal and private agendas to be advanced. Use of the models to help structure discussion enables us to do rather better than this.

Structure to the discussion is provided by using the models as a source of questions to ask about the situation. This phase of SSM has usually been referred to as a 'comparison' between situation and models, but this wording is truly dangerous if it is taken to imply that the discussion focuses on deficiencies in the situation when set against the 'perfect' models. *The models do not purport to be accounts of what we would wish the real world to be like*. They could not, since they are artificial devices based on a pure worldview, whereas human groups are always characterized by multiple conflicting worldviews (even within one individual!) which themselves change over time – sometimes slowly, sometimes remarkably quickly. (It is those conflicting worldviews which are the fundamental cause of the confusion in most 'management' discussion.)

No, the purposeful activity models simply enable our organized discussion to take place. From the model we can define a set of questions to ask. For example: 'Here is an activity in this model; does it exist in the real situation? Who does it? How? When? Who else could do it? How else could it be done?' . . . etc. Or: 'This activity in the model is dependent upon these other two activities; is it like this in the real situation?' There is no shortage of possible questions, and practitioners quickly develop the knack of passing in a light-footed way over many possibilities and resting on those questions which are likely to generate attention, excitement or emotion. The questions can be about activities or the dependence of one activity upon another or upon the measures of performance by which purposeful activity is judged.

A general finding is that groups find it very difficult to answer questions derived from the measures of performance in a model. 'What criteria would indicate the degree to which this activity (either individual, or the set of operational activities as a whole) is efficacious, efficient and effective?' This is usually a difficult question to answer in most real-world situations, due to their complexity, but it usefully draws attention to the need for organized processes of monitoring, something which is often given scant attention in organizations of all kinds. At a broader level, the fact that a given model is based upon a declared (pure) worldview will draw attention to other, usually implicit, worldviews which may underlie what is actually going on in the situation. This may serve to define other relevant models worth building and also helps to raise the level of discussion to that at which previously taken-as-given assumptions are now questioned. This will usually wake up anyone who is sleep-walking through the discussion, not least because differences of worldview always provoke *feelings*, not simply mental activity. (Also, incidentally, experience in developing SSM suggests that the stimulation of emotion is probably, for most people, a powerful trigger for significant learning to occur.)

In practice, several ways of conducting the questioning of the situation have emerged. An informal approach is to have a discussion about improving the situation in the presence of the models. If some relevant models are on flip charts on the wall, they can be referred to and brought into the discussion at appropriate moments. This has been found useful in situations in which detailed discussion of the SSM approach is inappropriate or is not feasible for cultural reasons. It was effective in a situation in a giant publishing/printing company which was characterized by an operation – publishing, printing and selling consumer magazines – which combined two very separate cultures who found it difficult to appreciate each other's worlds. The editor/publisher culture contained people very different from those in the printing culture, though they worked in the same company. Models which related to the whole operation of commissioning material, editing and assembling magazine issues, printing them and marketing them, proved useful here as a background, rather than as a source of specific detailed questioning. They were on flip charts on the wall, and could be referred to during discussion.

A more formal approach, probably the most commonly used, is to create a chart matrix as in Figure 2.12. The model provides the left-hand column, consisting of activities and connections from the model, while the other axis contains questions to ask about those elements (which may vary depending on the investigation underway). The task is then to fill in the matrix by answering the questions. Case 1 in Chapter 4 illustrates a question-matrix from an SSM investigation.

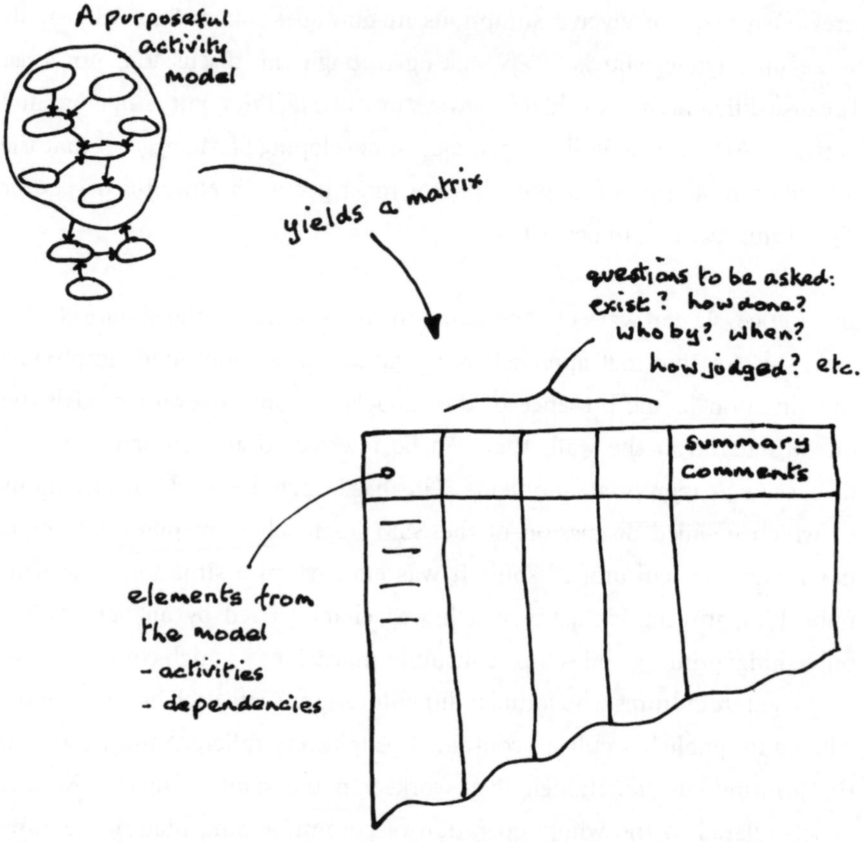

Figure 2.12 A formal process for using models to question the real-world situation

An important warning here is that this process should not be allowed to become mechanical drudgery. This is where a light-footed approach is needed, glancing quickly at many activities and questions, making judgements, and avoiding getting bogged down. Experience quickly develops this craft skill. In fact, experience suggests that this business of seeking to avoid plodding through every cell in the matrix itself helps develop insights into 'the real issues in this situation' – though such judgements have to be tested.

A third way of using models to question reality is to use a model as a basis for writing an account of how some purposeful action would be done

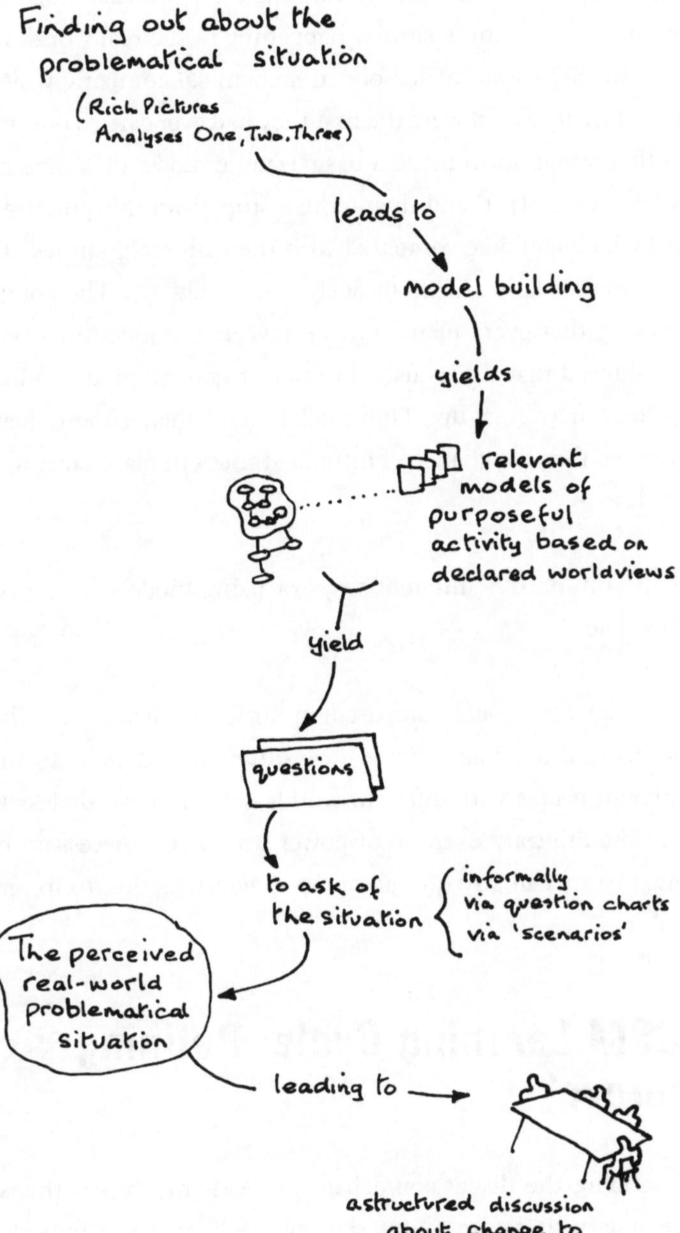

Figure 2.13 The role of models in SSM summarized

according to the model, and comparing this story, or scenario, with a real-world account of something similar happening in the real world. For example, work with SSM was carried out in a chemical company which treated every plant start-up as if it were the first they had ever carried out. It was very useful in that situation to make a basic generic model of 'a system to start up a new chemical plant' and then write a story from this pure (instrumental) model which could be compared with the real-world stories of previous plant start-ups, usually stories of delays and cock-ups. The company was right in saying that every plant start-up revealed unique features. But this work also showed that it was useful to have a generic model to hand when planning for a new start-up. This model could then be enriched by new experiences, so that the chance of future surprises in plant start-up could be diminished.

Figure 2.13 summarizes different ways of using models in the context of SSM as a whole.

Whichever way the models are used to structure discussion, the aim is the same: to find a version of the real situation and ways to improve it which different people with different worldviews can nevertheless *live with*. Outside of the arbitrary exercise of power, this is the necessary condition which must be met in any human group if agreed 'action to improve' is to be defined.

The SSM Learning Cycle: Defining 'Action to Improve'

When describing the discussion/debate in SSM, much – perhaps most – of the secondary literature about the approach makes a remarkable and fundamental error. It assumes that the purpose of the discussion/debate is to find *consensus*. It is a 'remarkable' mistake in that anyone who had read the primary literature with care would not make it, and it is 'fundamental' because, in order to cope with the complexity of human affairs, SSM uses a much more subtle idea than 'consensus'. It works with the idea of finding an *accommodation* among a group of people with a common concern. This does not abandon the possibility of consensus; rather it subsumes it in the more general idea of accommodation. A true consensus is the rare, special case among groups of people, and usually occurs only with respect to issues which are trivial or not contentious; issues which people do not feel particularly strongly about. In the general case, however, because individuals enter the world with different genetic dispositions and then have different experiences in the world, there will always be differences of opinion resulting from different worldviews. So, if a group of people are to achieve agreed corporate action in response to a problematical situation, they will have to find an accommodation. That is to say they will have to find a version of the situation which they can all live with. These accommodations will of course involve either compromise or some yielding of position. A compromise may give no member of the group all they personally would look for in action to improve the situation. But finding an accommodation is usually a necessary condition for moving to deciding 'what we will now do' in the situation.

The idea of finding accommodations is probably most familiar to us in our personal lives. Any family, as long as it is not of the classic Victorian kind, run by a (male) tyrant who decides everything, will have to continually find versions of the family situation which the different members can accept and live with. This is a necessary characteristic if families are to stick together over a long period. But the idea is also relevant to our professional lives, and to public life. A dramatic illustration of the latter is provided by some British political history. In the UK in the 1970s there were a number of major strikes in the coal industry, the disputes usually involving pay. One of those strikes lasted for a year. Now, the interesting thing about these disputes was that they were conducted within an accommodation between the two sides, the Coal Board and the National Union of Miners (NUM). Although the miners were on strike, members of the NUM nevertheless went down every mine in the country, every day, in order to keep the pumps running, since if you don't continuously pump water out of a coal mine you lose the mine. Although both sides regretted, but were prepared to have the dispute, there was an accommodation between them at a higher level: neither was prepared to live with the idea of the conflict destroying the whole industry. (It took political action to do that some years later!)

This view taken within SSM – that consensus is rare in human affairs, due to clashing worldviews – is not to be regretted. Clashing worldviews, always present, are a source of strong feelings, energy, motivation and creativity. If you find that the models you've built are not leading to *energetic* discussion, abandon them and formulate some more radical Root Definitions.

As discussion based on using models to question the problematical situation proceeds, worldviews will be surfaced, entrenched positions may shift, and possible accommodations may emerge. Any such accommodation will entail making changes to the situation, if it is to become less problematical, and discussion can begin to focus on finding some changes which are both arguably desirable and culturally feasible. In practical terms it is a good idea not to try and discuss the abstract idea 'accommodation' directly. It is best approached obliquely through considering what changes might be made in the situation and what consequences would follow. The relations between accommodations, consensus and changes is summarized in Figure 2.14, and the practical way forward in seeking accommodation is by exploring possible changes and noting reactions to them.

In doing this it is best to think richly about change in human situations, separating the concept into three parts for analytical purposes, even though any significant change in real situations will usually entail all three elements. These are: making changes to *structures*; changing *processes* or procedures; and changing *attitudes*.

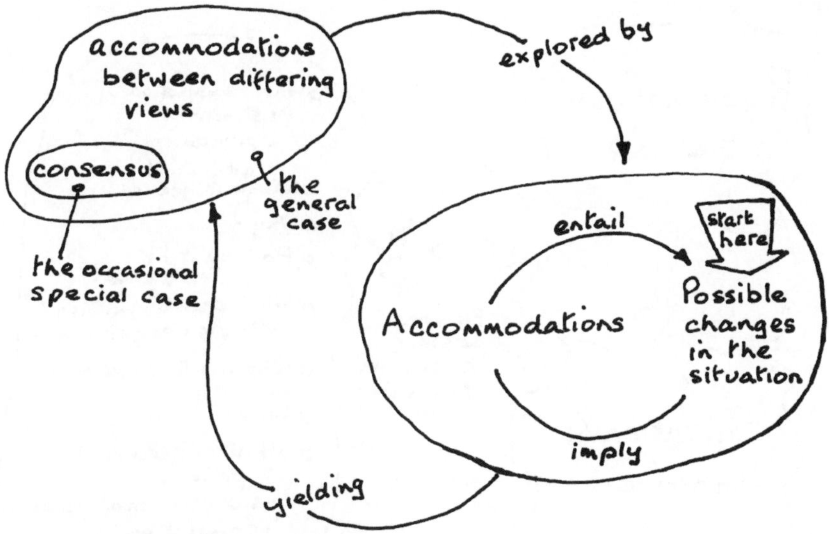

Figure 2.14 Seeking accommodations or (rarely) consensus by exploring implications of possible changes

Obviously the easiest element to change is structure, which can often be done by decree through the exercise of legitimate power. Researchers have noted, for example, that large organizations tend to reorganize themselves structurally about every 18 months to two years. In the UK, governments have imposed structural change upon the National Health Service more than 20 times since it was established in 1948. That is the easy part, for governments. But of course new structures usually require both new processes and new attitudes on the part of those carrying out the processes or being affected by them. Organizations (and governments) find it much harder to think out the necessary new processes; and no one can be sure, in a unique social situation, about what to do to change attitudes in a particular direction. (In our current culture, obsessed with economics, the usual mechanism for trying to change attitudes is to provide material incentives, but this reflects acceptance of a bleak model of human beings as creatures responding only to sticks and carrots. Human beings are more complex than that.)

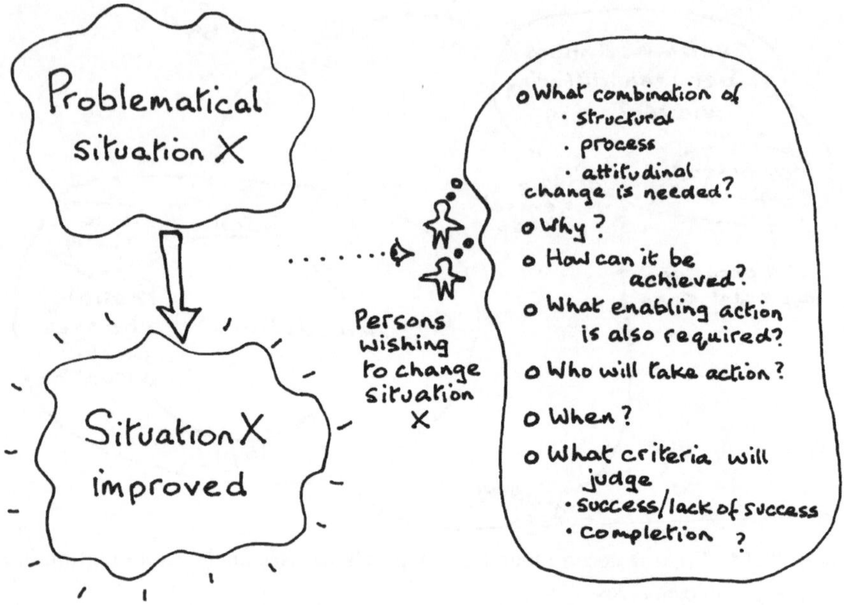

Figure 2.15 SSM's stance on introducing change in human situations

Figure 2.15 illustrates the stance on 'change' taken within SSM. It represents a reminder of things to think about when considering changes which are both desirable and feasible. It is self-explanatory, but two points are worth making. There is a question concerning the 'enabling action' which may be necessary if a potential change is to be accepted. This recognizes the social *context* in which any change will sit. Because of this context, introducing the change may require other action, enabling action, which is not directly part of the change itself. For example, when working within the UK National Health Service for the first time, in the early 1970s, the authors quickly found that in an acute hospital no proposed change would get accepted unless it had the support of senior hospital consultants. Shifts in the disposition of power have now modified that, but at that time in the history of the NHS, enabling action to secure the support of senior doctors was essential if any change of any kind was to occur in a hospital! The second point concerns trying to define the criteria by which a change can be judged as 'completed' and 'successful/unsuccessful'. This point has already been made above in connection with asking about 'monitor-and-control' activities in a real situation: well worth doing, but don't expect people in the situation to have any ready answers.

As we come to the end of this chapter's exposition of a 'fleshed-out' account of SSM, the discussion has become less detailed, in the sense that there are more detailed guidelines for finding out about a real-world situation, and building models used to question it, than there are for taking action to improve the situation. This is inevitable, and is due simply to the fact that no human situation is ever exactly the same as any other. Once we start exploring the real complexity of a human situation, not simply its logic, then formulae, algorithms and ready-made solutions are not available. Even guidelines become fewer. That being so, it seems helpful to give here, ahead of Part Two of this book, a real example of these ideas about change in action.

In the work mentioned above in the publishing–printing industry, the company carried out both of these major activities in selling a large range of consumer magazines. Publishing and printing were organizationally separate, and were in the hands of two very different cultures: on the one hand 'media-folk', on the other 'technologists'. There were many issues in the company concerning investment, pricing, and the placing and scheduling of work. For example, the printers thought of themselves as 'jobbing printers', making no distinction between printing one of the company's titles or that of a competitor. Publishers had ill-defined freedom to print within the company or externally. There were many rows about 'where to print', for example. This was an occasion in which the least-formal way of using models to question the situation was used: discussion in the presence of the models, which were on flip charts on the walls. In the discussion stimulated by the models the end point finally reached, subsequently approved by the board, was that there should be structural change. A new unit within the company was set up. This unit was centrally placed, and was staffed (part-time – it was not permanently in session) by people from both publishing and printing. This structural change was just about culturally feasible (where a fully integrated magazine-producing operation was out of the question) and the processes within the new unit were defined. As far as changes of attitude were concerned, the chief executive, who understood the difficulties of forcing change of that kind, wrote in the in-house company 'newspaper': 'Primarily the new unit is concerned with trying to develop a more effective relationship between our publishers and printers.' He was hoping that each of the two cultures would, through working together on some issues, begin to see the world through the eyes of the other.

The Whole SSM Learning Cycle Revisited: Seven Principles, Five Actions

Before moving on to accounts of SSM in action, in Part Two, we can now summarize the whole learning cycle of the SSM approach. In a concise account of SSM, which is as spare as we can make it, seven principles lead to five actions. These are based only on findings which, through many experiences over a long period, always turned out to be helpful. They are the end product of the several hundred cycles through the LUMAS model in Chapter 1 (Figure 1.8).

The seven principles which underlie SSM are set out first.

1. The idea 'real-world problem' is subsumed in the broader concept of 'real-world *problematical situation*'; that is to say, a real situation which someone thinks needs attention and action.

2. All thinking and talking about problematical situations will be conditioned by the *worldviews (Weltanschauungen)* of the people doing the thinking and talking. These worldviews are the internalized taken-as-given assumptions which cause us to see and interpret the world in a particular way (one observer's 'terrorism' being another's 'freedom fighting').

3. Every real-world problematical situation will contain people trying to act *purposefully*, with intent. This means that *models of purposeful activity*, in the form of systems models built to express a particular worldview, can be used as *devices* to explore the qualities and characteristics of any problematical human situation.

4. *Discussion and debate* about such a situation can be *structured* by using the models in (3) as a source of questions to ask about the situation.

5. Acting to improve a real-world situation entails finding, in the course of the discussion/debate in (4), *accommodations* among different worldviews. An accommodation entails finding a version of the situation addressed which different people, with different worldviews, can nevertheless live with.

6. The *inquiry* created by principles (1) to (5) is in principle a *never-ending process of learning*. It is never-ending since taking action to improve the situation will change its characteristics. It becomes a new (less problematical) situation, and the process in (3), (4) and (5) could begin again. Learning is never finished!

7. Explicit organization of the process which embodies principles (1) to (6) enables and embodies *conscious critical reflection* about both the situation itself and also about the thinking about it. This reflection, which leads to learning, can (and should) take place prior to, during and after intervening in the situation in order to improve it. The process thus itself virtually ensures *reflective practice* by those who make use of it. Once the practitioner has internalized the SSM process, so that he or she no longer has to stop and ask questions about it ('Remind me again, what did PQR stand for?') then reflective practice becomes built-in too. The SSM user becomes a reflective practitioner.

These seven principles clearly underlie the four actions which define the classic shape of SSM in Figure 1.5: finding out about a problematical situation;

making models relevant to exploring it, based on different worldviews; questioning the situation using the models, in order to find desirable and feasible change; and defining/taking action to change the situation for the better. The seventh principle itself defines a fifth action which ensures cycling round the primary four, namely critical reflection on the whole process. This fifth action is at a different level from the other four. It is *about* the other four, i.e. at a meta-level. It is the activity which ensures that the lessons learned are captured, in the way that the LUMAS model of Figure 1.8 indicates. Figure 2.16 expresses these five activities at their two levels.

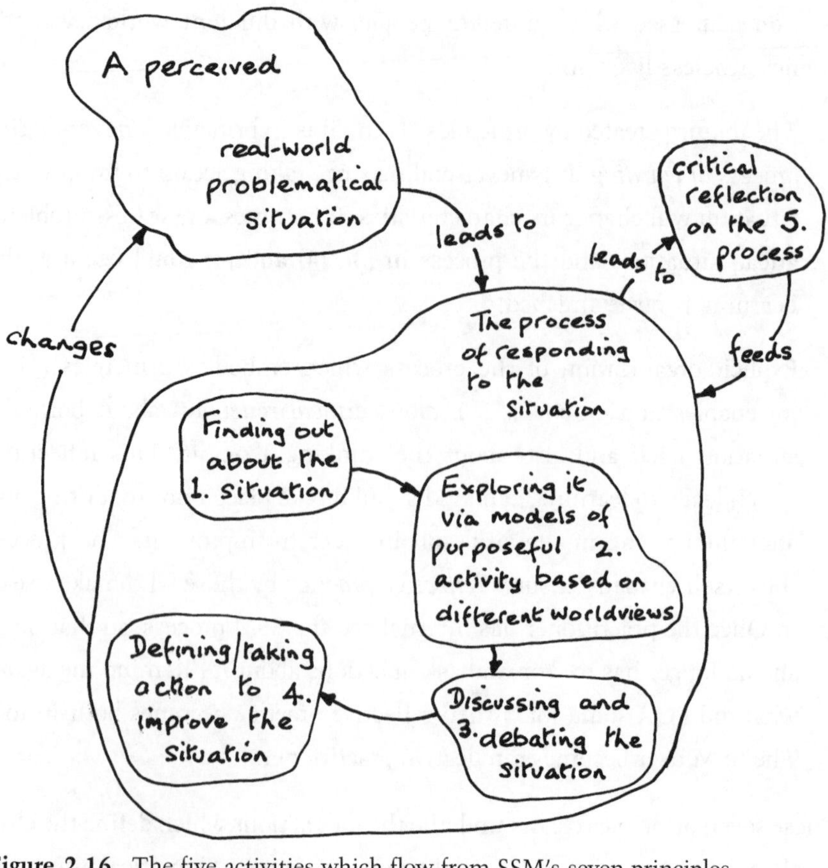

Figure 2.16 The five activities which flow from SSM's seven principles

Finally, in completing this more detailed account of SSM, it is worth re-emphasizing some of its core ideas. It does not seek 'solutions' which 'solve' real-world problems. Those ideas are a mirage when faced with real-life complexity, with its multiple perceptions and agendas. Instead SSM focuses on the process of engaging with that complexity. It offers an organized process of thinking which enables a group of people to learn their way to taking 'action to improve'; and it does that by means of a well-defined, explicit process which makes it possible to recover the course of the thinking which leads to action. This makes sure that every use of the approach produces learning which will accumulate over time, leaving the user better equipped to cope with future complexities.

Part Two now provides examples of these ideas in action in situations of many different kinds.

Part two
SSM in action

3
SSM in Action in Management Situations

Introduction

It will be obvious from the previous chapters that in the development of SSM there was continuous interaction between rich experience in real situations and thinking about the emerging methodology. In Chapters 1 and 2, with the focus only on describing the mature methodology, we have had to separate experience from methodology. This means that these chapters are inevitably rather austere. Now in Part Two we can restore some of the richness by relating the methodology to real situations in which it has been used. We can show the variety of its use, in small companies and large corporations, in the public sector and in private organizations. We can describe investigations at different levels, from tactical to strategic, and we can include investigations which took a few hours as well as those carried out over many months.

In doing this we hope in each case to illustrate how the methodology, as a set of underlying principles and five basic activities, is always tailored to meet the requirements of the particular circumstances of the situation being addressed. Just as every human situation is in some respects unique, so no two ways of addressing the five activities of SSM are ever exactly the same.

Given this emphasis on the tailoring of the methodology activities to particular situations, the accounts will be kept short deliberately, so that a range of different kinds of situation can be covered. For each story a reference indicates where a more detailed account can be found. The shape of each of these short accounts is as follows: a description of the situation addressed, how SSM was used, and the outcomes.

Part Two has three chapters. This one contains accounts of SSM in action in a range of problematical situations in the broad field of management; Chapter 4 collects examples in which the area of application is the more specific field of information systems. Chapter 5 is rather different. Although the ideas incorporated in SSM are not fundamentally difficult to understand, deeply appreciating the approach does entail, for most people, some shifting of mental furniture, which is rarely an easy or comfortable process. That chapter describes common misconceptions of SSM and mistakes made in describing it. The aim there is to dispel confusion and make possible a deeper level of understanding.

Case 1 Rethinking the Role of a Head-Office Function in Shell

The Situation

In every large corporation there are departments or divisions which are 'front-line' in the sense that they actually carry out the activities which directly generate wealth. And there are supporting 'head-office' departments which oversee the activity of the whole and carry out activities which try to ensure that the corporation generates more wealth from 'front-line' activity than it uses up in doing all of its activity – which is a necessary condition for the continued existence of any organization that is not subsidized and so has to earn its keep. Well-managed corporations will try to ensure that the head-office functions are both lean, since they do not directly generate wealth, and effective in helping to achieve long-term aims. Inevitably there will be tension between front-line and head-office departments, with people in the former wondering what the people in head office find to do all day. The trick is to make that tension creative rather than destructive.

Case 1 was an SSM-based investigation into a head-office department in the Shell Group, based in its headquarters in The Hague. The department was the manufacturing function (MF) with 600 people whose long-term mission, fundamentally, was to ensure that Shell's technology and its research and development activity were at least as good as any in the global oil industry. The investigation came about when a new head of MF, Rob de Vos, decided that a root-and-branch rethink of MF, to include its role, structure and procedures, should be carried out. Having consulted one of his chief planners in MF, Jaap Leemhuis, and also Kees van der Heiden from the Planning Group in Shell's London head office, he accepted their idea that a systems study using SSM should be mounted. Checkland was asked to help with the facilitation of this, they being clear from the start that they sought a broad participative investigation by many Shell managers. This would include not only MF-ers, but also some who dealt with MF as 'customers', as well as other Shell professionals. It was impressive that Rob de Vos was prepared to allow a 're-orientation of MF' (his phrase) to emerge in the SSM-based study; he did not try to direct it, and was prepared to contribute to it himself as a participant. This presented a fine opportunity to use mature (late 1980s) SSM in a situation of significant importance and complexity.

The Use of SSM

Prior to the SSM-based study, Leemhuis and van der Heiden had carried out an extensive 'finding-out' process consisting of 80 interviews with managers in MF and other control functions, and also with 'customers' of MF both inside and outside Shell with whom MF interacted. Findings were fed back to workshops of MF managers in order to try to create a shared perception of MF's role and the issues it faced as it carried out many activities at different levels, over different timescales, and with many customer-clients.

The picture that emerged was complex, with MF seen as being required to live with many conflicting requirements:

- focus on technology development, but provide broad-based advice to very many internal customers;
- employ generalists with a deep understanding of the business context, but develop 'old foxes', technologists with the understanding which comes from years of technological practice;
- keep managers in post for long enough to deeply understand MF problems, but rotate managers regularly so that they keep in touch with the realities of life in the refineries, etc.

This provided (in more than 100 pages) a compendium of Shell opinion concerning MF's role and the issues surrounding it.

In starting the SSM investigation the three facilitators (Leemhuis, van der Heiden, Checkland) decided to make use of a common Shell practice: the organization of two-day management workshops, held in hotels, in which 15 to 20 selected managers discussed a current issue. Findings or conclusions would be collected in a report which was circulated to a wider audience than the participants of the workshops. Methodologically, it was agreed that the workshops would be conducted in the everyday language of Shell managers, rather than SSM language, but that the process being followed would be explained if questions were raised about it. Each workshop would have a theme, and the theme of the next workshop (with mainly new participants but some overlap) would be decided during a six- to eight-week gap between workshops. In planning this investigation SSM (p) was not used formally but Checkland made the model in Figure 3.1 as an aide-memoire for himself.

Similarly there was no question of persuading managers coming to a seminar with what they saw as a specific task (namely, contributing to the 're-orientation of MF') to spend time discussing Shell's culture and politics. Instead Checkland, attending the seminars and working to shape their

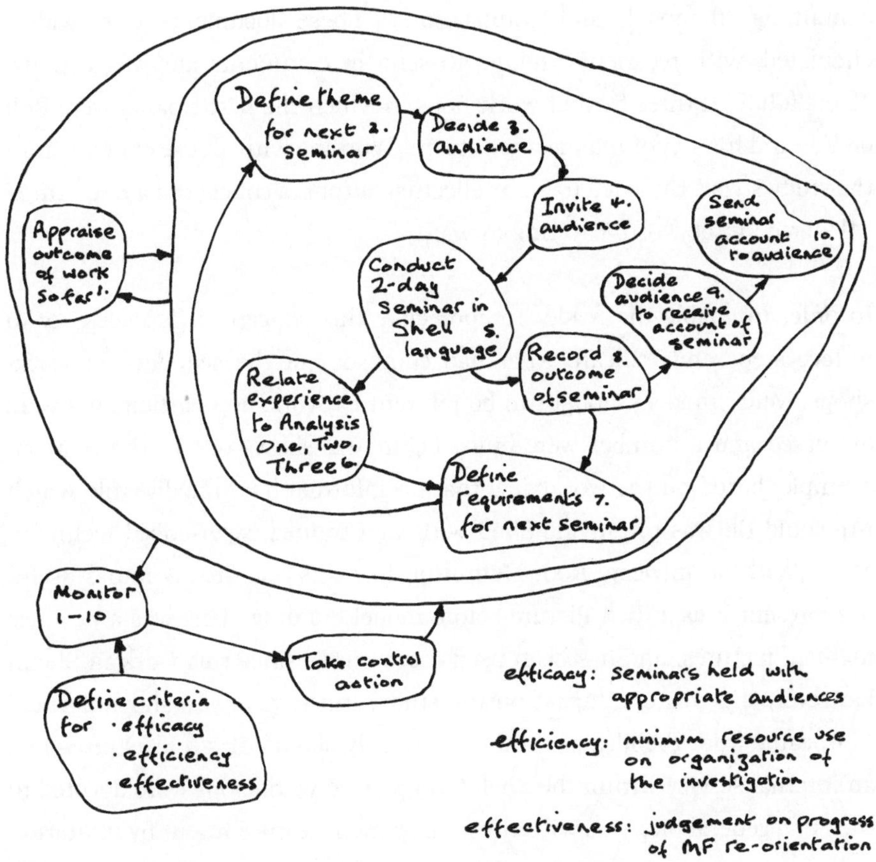

Figure 3.1 The SSM (p) model from the rethink of MF in Shell

content, used the experience to compile his own files on Analysis Two and Three, so that he was better prepared when it came to understanding feasible 'action to improve' MF later on.

Following the initial exploration of issues by Leemhuis and van der Heiden, five two-day seminars were held. The first two (with themes 'technology development' and 'a service business') set the pattern: day one – discussion followed by model building; day two – using the models to question the real world of MF in Shell. Each seminar led to a document of 40 to 50 pages containing all models and 'comparisons'. These documents were widely circulated, with recipients invited to send in comments and suggestions. There followed three further workshops in which the participants were Rob de Vos and his senior management team. Working with all the content from the study so far, these led to an intellectual output: a concept of a re-oriented MF agreed to be worth carrying forward.

In order to enable the reader to appreciate this concept, it is necessary to retrace steps and indicate what had come out of the sequence of workshops. Many models thought to be relevant by workshop participants were discussed and a number were built. Figure 3.2 shows one of these, as an example, based on the worldview that an informal but valuable role which MF could fill was to provide Shell with well-trained experienced technologists. (With an intricate Root Definition like this one, it was found useful to represent it as a Rich Picture before model building. This was a new use for such pictures, and has been used many times since this work in Shell.) However, the greatest impact on the study, in the two workshops devoted to it, came from the idea of thinking seriously about MF's role as providing an internal service within the Shell Group. Two workshops were devoted to this, one general one, one with the management team. Present by invitation at both of these workshops was Richard Normann, a management consultant who has written extensively on the need for *every* business to think of itself as providing a service. His contributions fed discussion on seeing part of MF's role as providing a complex mix of services within the Shell Group. A concept of service provision which found favour with workshop participants is that shown in Figure 3.3.

In order to make this abstract concept more useable, it was expressed in the root definition and activity model shown in Figure 3.4. This is of course still highly abstract, but by seeing MF as Party A in Figure 3.4 and placing various groups in the role of Party B (e.g. Shell Main Board, refineries requiring day-to-day technical support, partners in joint projects, etc.) it was possible to

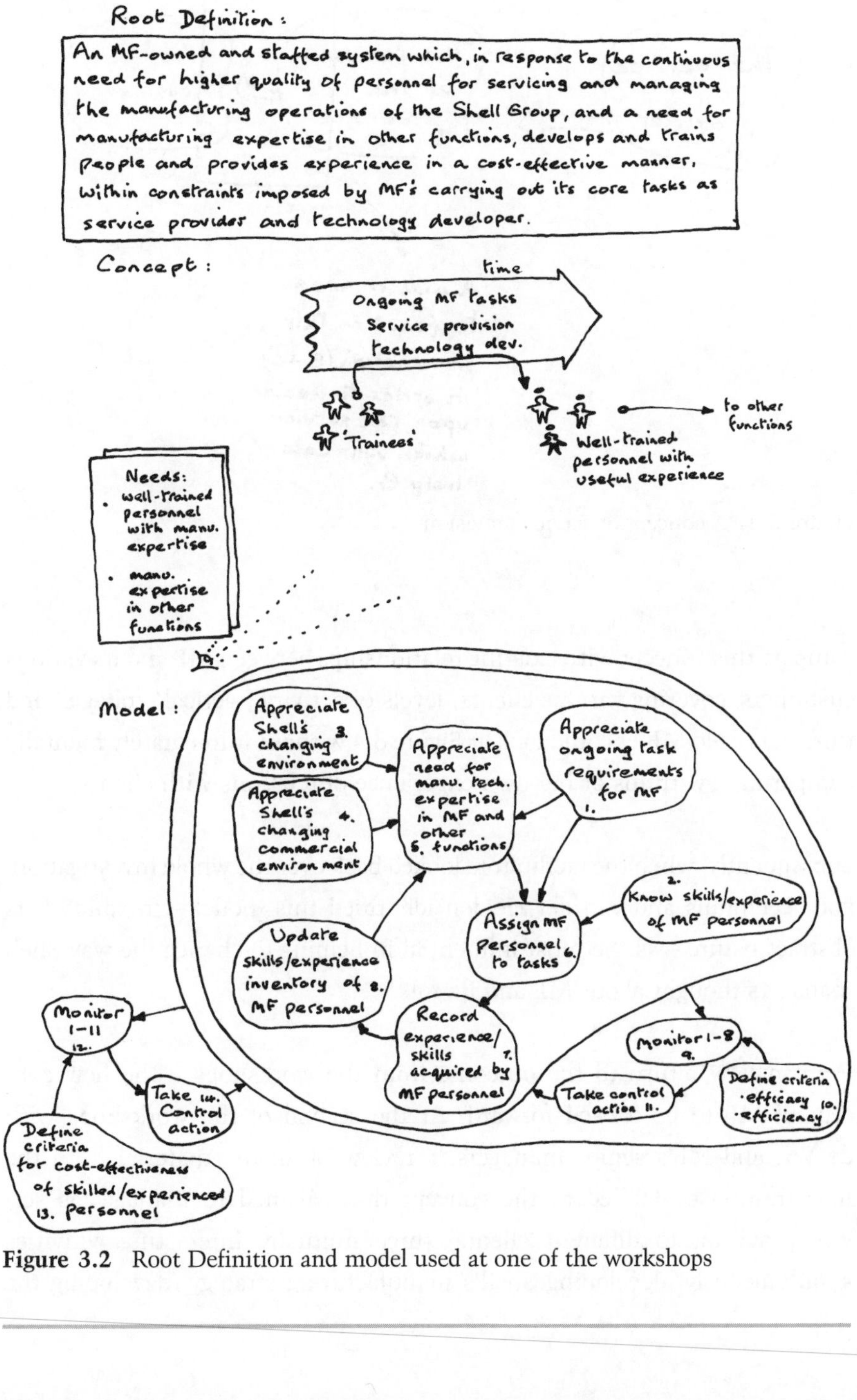

Figure 3.2 Root Definition and model used at one of the workshops

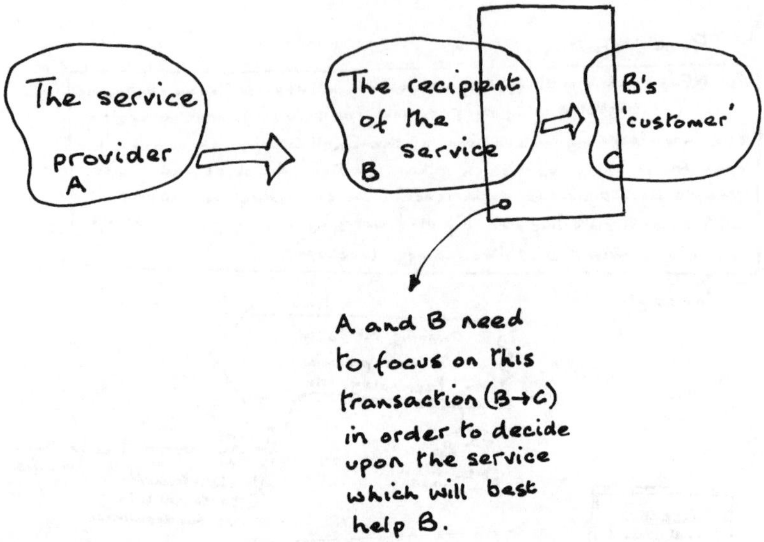
The service provider A
The recipient of the service B
B's 'customer' C
A and B need to focus on this transaction (B→C) in order to decide upon the service which will best help B.

Figure 3.3 A concept of service provision

compare this concept with existing relationships between MF and its various customers, covering various clients, levels of activity (tactical/strategic) and timescales. No MF-er could look at Figure 3.4 without immediately mentally comparing it with his or her own experience of relations with clients.

Subsequently, when the facilitators looked back over the whole investigation, both Leemhuis and van der Heiden identified this model – in spite of its abstract nature – as the most influential in helping to change the way Shell managers thought about MF and its role.

We can now return to the outcome from the workshops – the new concept of MF to be carried forward. At the second of the workshops with de Vos and MF's senior managers, a review of all the material from the investigation so far led to the concept that MF had to manage (as service provision to different clients) three mutually interacting activities simultaneously: developing Shell's manufacturing strategy, developing the

as devices from the world of logic, rather than descriptions of messy reality.) This workshop was recorded in a 43-page report, and this led to discussions and consultations over a month or so, much of it devoted to the way in which feasible new structures and processes for MF could be made public, explained, argued and made to work.

At the end of this period, Rob de Vos addressed more than 500 members of MF, describing a new structure for MF and enlisting their help in making it a reality. The new structure consisted of new organizational units (human resource planning and development, support services, technology development, strategy development) each perceived in terms of running a service business for a named client group, with both activities and measures of performance named, as in Figure 3.8.

In his address he also indicated that several seminars for 40 people at a time, chosen from many levels across the MF function, would be organized to enable MF-ers to 'express concerns and generate ideas for the further development of new ways in which we can conduct our activities'. Several 'lunchtime information sessions' for 120 people were also organized. The seminars-for-40 generated 2,355 comments and suggestions (!), which Leemhuis and Van der Heiden analysed and grouped in a 65-page book which went to every member of MF. In the end, every member of the function who wanted to had the chance to contribute ideas and suggestions concerning the new MF. This made the 'Re-orientation of MF' a remarkable and unusually open programme of organizational change, very different from the more usual chaotic and rambling discussion which gets ended by some senior management decree.

The whole process of the investigation is shown in Figure 3.9. It was, of course, a part-time activity for all concerned, and was spread over a period of 15 months from the initial gathering of perceptions of MF to the new MF in operation. This allowed time for the new thinking to be absorbed.

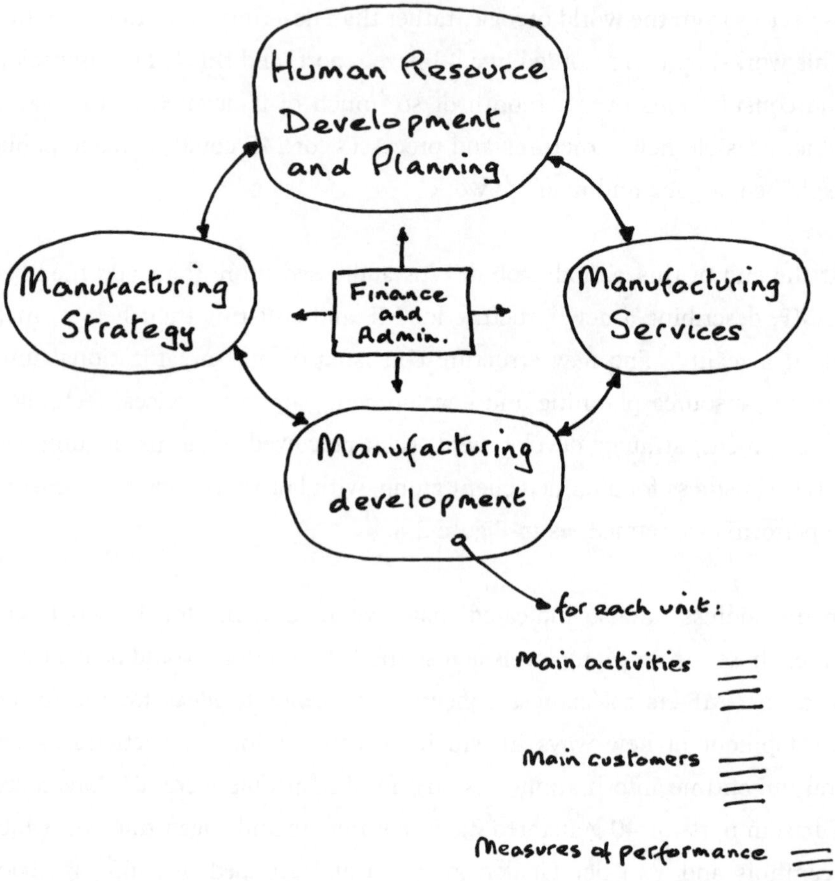

Figure 3.8 The form in which the new organization structure of MF was expressed

Outcomes

There is no need to dwell on the practical outcomes. The SSM-based project led to a new way of thinking about MF, a new structure and new processes. This implied new attitudes by MF-ers, and the elaborate work done after the third management workshop was aimed at the necessary attitudinal change. This seemed to be acceptable to most but there was an interesting exception. One senior manager in MF summarized the whole intervention in the following terms: 'Though we could not have articulated it at the start, we

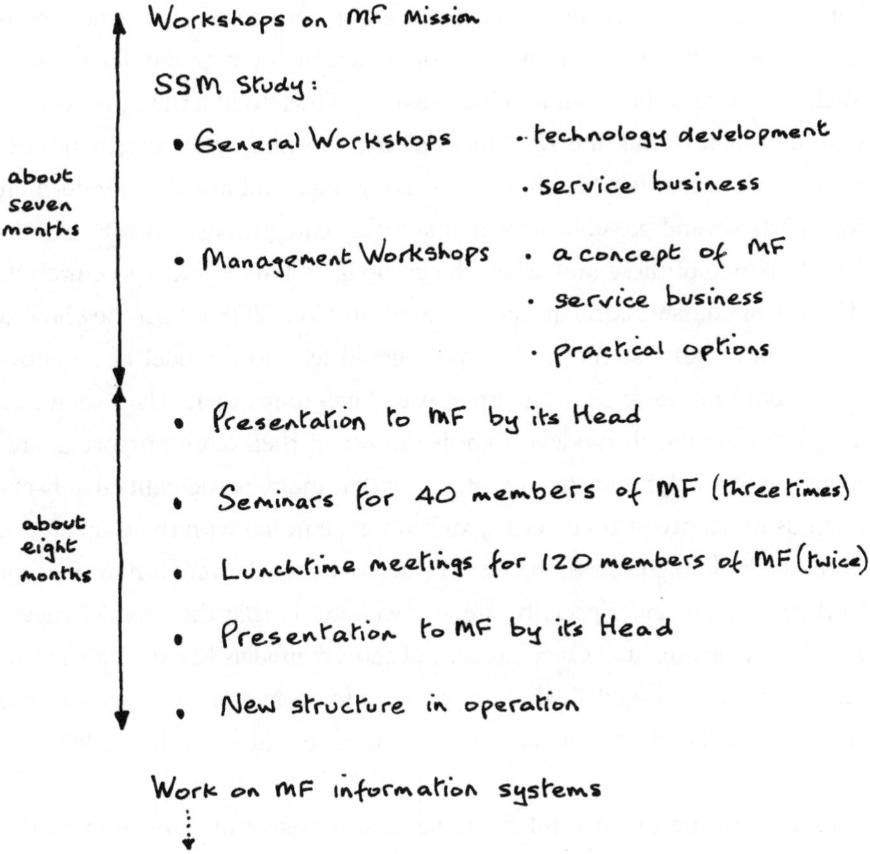

Figure 3.9 The whole process of the 'MF re-orientation'

were at that stage thinking of MF as a creator and preserver of technical skill pools. Now we are thinking of it as an internal service business.' This was a very neat summary of the change of worldview, but for the manager who said it, the shift was too much. He saw himself as a professional technocrat, and wished to remain that. He asked for, and got, a move out of MF rather than accept a new concept which for him had a whiff of commerce about it.

In terms of methodology use, SSM was the crucial means for structuring a long and complex investigation. It prevented discussion becoming circular – which happens very easily in most undirected groups of people – and was unobtrusive since the investigation was conducted in the everyday language of Shell managers rather than in SSM language. Thus, to illustrate this, on the first afternoon of the first workshop, after a morning spent in general discussion, participants were split into small groups and asked to produce on flip charts several possible statements of the 'core purpose' of MF. Checkland took one of these and asked the group questions in order to enrich it. (He was of course asking questions based on CATWOE.) Then he showed how such a rich statement of purpose could lead to a model of an activity system to achieve the stated purpose. The groups themselves then had a go at making such models from a number of their 'core purpose' statements. It was helpful in this investigation that many participants had backgrounds in science or engineering, and so were familiar with the general idea of model building, even though SSM's activity models were a new concept to them. (Engineers, especially, are also well aware that their models never match complex reality.) Once the idea of activity models based on a particular worldview was established, there was a readiness by workshop participants to relate to and make use of 'ready-made' models produced by the facilitators.

This investigation also reinforced the lesson concerning the importance of recording and disseminating what goes on within an organized study. It will be obvious from this summary that Shell were extremely good at this. The widely-distributed workshops reports, the compilation of ideas and suggestions from members of MF, etc. all made it clear that an open and transparent process was being followed. And the distribution of these documents made it possible for very many people to feel that they knew what was going on, and could contribute.

Finally, the idea of expressing complicated Root Definitions as Rich Pictures – as a way of easing passage to model building – was an innovation in SSM use which has now become part of the approach when elaborate definitions are being used.

A much more detailed account of this investigation is given in Chapter 9 of *Soft Systems Methodology in Action* by Peter Checkland and Jim Scholes (John Wiley & Sons, Ltd, 1990).

Case 2: Evaluating Past Events

The Situation

This short illustration of the use of SSM offers a considerable contrast to the 15 months of work in a multi-national corporation that has been described in Case 1. This investigation entailed four weeks of work, spread over three months, in a small organization. The systems thinking involved occupied only half a day of the four weeks, but this experience contributed significantly to the development of SSM (as well as being useful to the client).

The client, here called Adam Cliff, was a mining engineer. After many years of experience, and with many contacts across the whole industry, he had formed his own management consultancy to work in the global mining industry. He had recently carried out a study for one of his clients, who manufactured the hydraulic supports which keep floor and roof apart in a mine ('Mining Support Ltd'). Unfortunately he had to regard this piece of consultancy work as a complete failure. Wishing to learn from this bad experience, he approached ISCOL Ltd, the consultancy wholly owned by Lancaster University which traded for 20 years, and was set up by its postgraduate Department of Systems in order to gain access to real-world problematical situations. Adam Cliff asked for a retrospective analysis of his unsatisfactory experience aimed at finding out what went wrong. His reason for approaching ISCOL was that he was strongly of the view that commercial mining operations always had to be approached using 'a systems approach'. What he meant by that was that the essence of any working mine ought to be thought of as 'a total system', as shown in Figure 3.10.

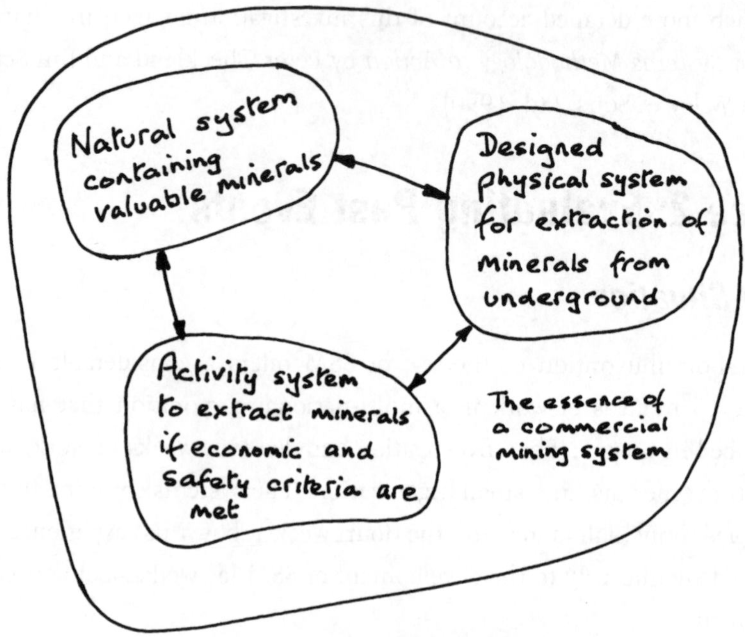

Figure 3.10 Adam Cliff's concept of the essence of any working mine as a system

Here three sub-systems are always linked: a natural system containing the mineral to be extracted, a designed physical system (shafts, roadways, extraction equipment, hydraulic supports, etc.) which is the mine, and an operational (activity) system to carry out the extraction as long as it was safe and economic to do so. No one who thought systems ideas powerful would want to argue against this way of thinking about mining, though in this particular piece of work it turned out to be not very relevant.

In order that retrospective analysis of the interaction with Mining Support Ltd could be carried out, Mr Cliff provided access to his staff and to all the files on the study in question. He also arranged similar access to his client, who were surprised but ready to collaborate. For ISCOL the analysis was undertaken by Peter Checkland and David Brown, the latter being at that time General Manager of ISCOL Ltd. They surmised that Cliff's motivation in setting up the ISCOL study was twofold. He was certainly puzzled as to

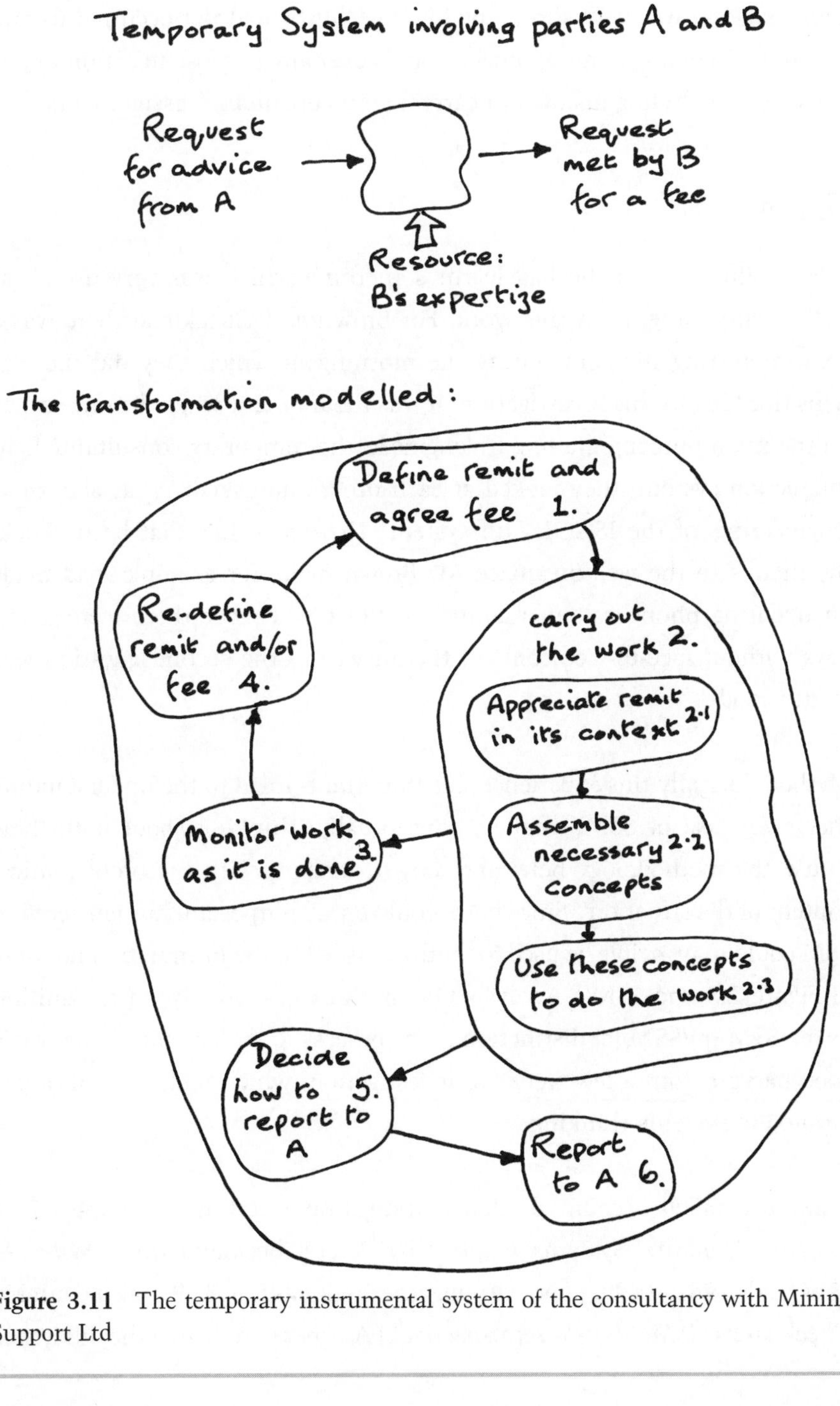

Figure 3.11 The temporary instrumental system of the consultancy with Mining Support Ltd

This outcome was very surprising for Mr Cliff. It had not occurred to him that the idea of a systems approach could cover any purposeful action in real situations, including his own in carrying out consultancy assignments.

Outcomes

Adam Cliff said that he had learnt something which was very useful, as well as surprising, from this work. For Brown and Checkland there was a very interesting moment during the morning in which they did the systems thinking in this investigation. It was the moment when, having settled on the main outcome (failing to 'engineer' the temporary consultant/client interaction system) they looked at each other and asked: What about *our* 'engineering' of the ISCOL/Cliff system? How good has that been? Luckily, thanks to the well-organized Mr Brown (who, for example, had made an updating phone call to maintain contact whenever there were a few days without face-to-face contact) the answer to the second question was 'pretty good'.

Methodologically this experience was important; it led to the understanding that *every* use of SSM entailed, conceptually, thinking about both 'how to use the methodology here' and 'how to grapple with the problematical content of the situation'. Since both would entail purposeful human activity, SSM could in principle be used for both. This led to the formulation not only of Figure 2.3 and SSM's Analysis One but also, eventually, to recognition of the SSM (p)/SSM (c) distinction (p for process; c for content). This was a good harvest from a few weeks of investigation, which included half a day devoted to systems thinking.

A more detailed account of this investigation is given in Chapter 7 of *Systems Thinking, Systems Practice* by Peter Checkland (John Wiley & Sons, Ltd, 1981, 1999. (The current paperback edition, 1999, also contains Checkland's '*SSM: 30-year retrospective*'.) A paper concerning the difference between SSM (p) and SSM (c) (by Peter Checkland and M. Winter) will be published in a special issue of the *Journal of the Operational Research Society* on 'Problem Structuring Methods' during 2006.

Case 3: Upgrading the Skills of a Department

The Situation

This use of SSM came about when a departmental manager in a large science-based manufacturing company, 'Regal Chemicals Group', attended a one-week course on SSM. He decided that the SSM approach could help him achieve one of his current aims, namely to up-grade the problem-tackling skills of his department. Approaching ISCOL Ltd at Lancaster University, he asked for help which would enable three of his staff to carry out an SSM-based investigation, aimed at achieving a broader outlook and the enhanced skills which would go with that. ISCOL assigned Peter Checkland and ISCOL consultant Iain Perring to the task of helping the team of three from the department, who were given a guaranteed one-day-a-week to work on the investigation over however many weeks the work took.

The department in question was a staff function known as the Information and Library Services Department (ILSD). It was located at the company headquarters. ILSD had several responsibilities: managing and growing the company library; meeting requests for information from company employees; managing Regal's involvement with computerized global information systems and databases; maintaining a classified, retrievable collection of specimens of every chemical compound (more than 50,000 of them) made by Regal researchers; and managing access to a secure collection of confidential internal company reports with defined restrictions as to who was allowed to read them. The chosen team had skills and experience across this range of activities, but they had never carried out anything like the SSM-based investigation, and had no knowledge of SSM.

The study they carried out is a classic use of the whole methodology, and exhibits four cycles of methodology use.

The Use of SSM

Work began with a two-day meeting. This served to introduce the team to SSM and allay fears about it, and to introduce Checkland and Perring to the current situation in the company. SSM was introduced to the team using Regal material relevant to the team's investigation, rather than theoretically, and Analysis One (of the intervention itself) was completed first. The 'client' was the Head of Department (HoD); the issues were being tackled by the three ILSD managers chosen by the HoD, together with the two facilitators; and the possible 'issue owners' included potential and actual users of ILSD services, ILSD professionals, the company itself, and the HoD.

Attempts to get the ILSD team to carry out Analyses Two and Three formally were not very successful. This is a common finding. People living in the situation under investigation usually feel that they know it so well that a formal examination of culture and politics is not necessary. In fact, of course, what they 'know' is whatever becomes taken as given in the organization from day to day; it is the *unexamined* features of their daily lives. Accepting SSM's modest disciplines in Analyses Two and Three can yield great insights and can lead to new perspectives and new motivations for change. But facilitators often have to be patient, stimulating cultural and political analysis over a period of time by asking pertinent questions. This is what happened in Regal. It gradually emerged that the ILSD team saw their roles and norms in terms of responding quickly and effectively to requests from users of their services by exercising professional skills. They accepted that they needed to maintain open, welcoming links with enquirers; and they recognized very clearly that though they had a commodity of power resulting from their possessing professional knowledge, there were key individuals within the company who, regardless of their status in the management hierarchy, were informal 'gatekeepers' affecting ILSD's relationship with particular departments or groups. This led to remarks of the kind: 'If you want to influence the market researchers, make sure Tom Smith is on your side, even though he's not Section Head.'

The completed Analysis One was fruitful in leading to potential 'relevant systems' to model, but presented a problem in that 26 possibilities were quickly generated. The first cycle through the methodology tackled the task of reducing this number. The 26-strong list contained such ideas as:

- improve liaison with, and feedback from users;
- decide how to define the acquisition, retention and discard policy;
- appreciate, absorb and exploit technological development.

All 26 ideas were considered 'relevant' but there was no wish to define 26 Root Definitions and build 26 models. Progress was made by putting the 26 possibilities to one side, temporarily, and conceptualizing, at a higher level, ILSD's position in Regal Chemicals Ltd. This was captured in the simple concept shown in Figure 3.12 – the EROS model.

ILSD can be thought of as one of a number of enabling systems which support Regal's wealth-generating operations. There was no need to convert Figure 3.12 into an activity model. Each of the 26 possibilities was 'compared' with the model in order to define its focus: was it about E, R, O or S? Doing this enabled the ILSD team to group their 26 ideas, and to form views on priorities. The debate structured by this 'comparison' lifted the level of thinking, and led to a choice of four ideas to carry forward, all at a somewhat higher level of abstraction than the 26:

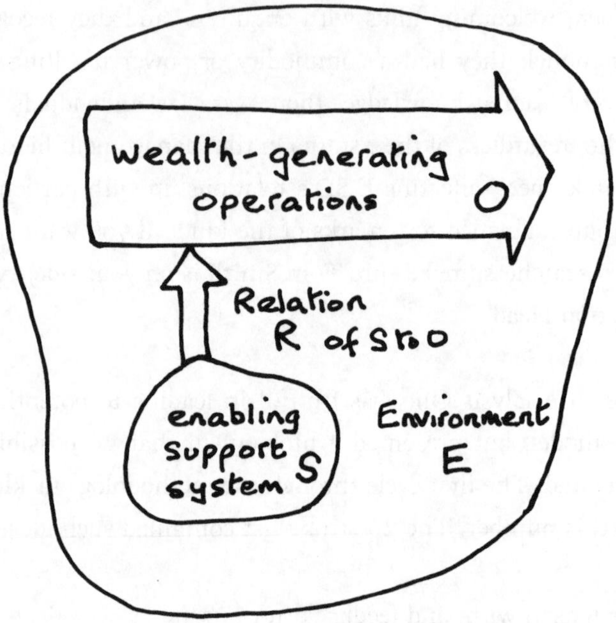

Figure 3.12 The rudimentary EROS model of operations and enabling support

1. an environment-appreciating system;
2. a relationship (O–S) establishing and maintaining system;
3. an 'information-as-a-resource' system;
4. an aiding-the-business system.

In a second cycle of the methodology, the third possibility was modelled first, since the other three were seen as ancillary to it. Two Root Definitions were based on relevant system (3).

RD3a

A system for the Regal board to manage an information resource, involving the creation and manipulation of a database from locally generated and bought-in material, covering the company's present and possible future research and business activities as an aid to decision making, within manpower and financial constraints

C Regal board and company
A information professionals
T need for a managed information resource → that need met
W this kind of staff support function is needed and is feasible
O Regal board
E structure of line/staff functions, manpower and financial constraints.

RD3b

A Regal-owned system for satisfying the potential and stated information needs of the user by the timely provision of a comprehensive, readily available collection of information, together with a facility/service to meet information needs.

C the user
A information professionals
T user with information needs → user with those needs met
W it is possible to define and provide for users' information needs in Regal
O the company
E staff and line functions; various conditions: timely, comprehensive, etc.

Models from these Root Definitions initiated about 20 days of work in which detailed tabular comparisons were carried out and further specific models were built. To the facilitators this period indicated a shift in perception as the ILSD team gradually became comfortable with a previously alarming idea: that of not only responding quickly and effectively to requests, but also of being in a position to tell actual or potential users things they *ought* to know.

In a third cycle of the methodology, one of the more specific models from the previous cycle (shown in Figure 3.13) had been found to be particularly useful, and it became a focus. Activities 4 and 5 in that model were then

Figure 3.13 Root Definition and model for an information provision system from the providers' point of view

themselves modelled in more detail. The root definition and CATWOE used to expand these two activities in Figure 3.13 were as follows.

> A system owned by ILSD which, together with users, identifies those scientific, technical and commercial staff in research, technical service,

development, production and business functions who require professionally provided information to do their jobs effectively, and key users in particular; and which identifies the level and nature of those requirements, i.e. breadth and depth of subject matter, detail and precision of output.

C ILSD and users
A ILSD and users
T identification of users and users' needs
W this mutual working can be an effective and efficient basis for ILSD processes
O ILSD
E existing organization structure

(The 'key users' mentioned in this RD are the 'natural gatekeepers' identified in Analysis Three.)

This increase in the level of detail indicated a move by the team from thinking about *what* ILSD might do to *how* it might do it; and this was allied to an overall shift from thinking of ILSD as a reactive provider of information, to thinking of it as a proactive provider working with its users.

By this stage of the work, the ILSD team were comfortable with the idea of a more sophisticated role for the department. Since they were not themselves in a position to implement it, they now needed to widen the audience for this work within the company. With the agreement of their HoD, who had kept closely in touch with the work without interfering in it, they made a presentation to all members of their own department and discussed its implications. They also wrote a detailed report on the work aimed at Regal's senior management. Significantly, and it is a measure of their increased confidence, the team of three did not wish to share authorship of the report with the two facilitators, and it was the team who made the presentation to management which followed internal distribution of the report.

The report presented ILSD as being a provider of:

> A continuing comprehensive information service to users, employing technical and other skills and modern technology so that the service is effective in furthering Regal's business interests.

This concept was then compared with current arrangements and four proposals for change were made

1. a more central role for ILSD in conjunction with its users;
2. a relationship with 'active' users (trained by ILSD to help themselves) and a richer user–ILSD dialogue;
3. a training function for ILSD;
4. a more organized monitoring and control of ILSD activity.

The presentation to management went well. A senior manager from the Research Department said to Checkland after the presentation: 'I've known and worked with ILSD for 20 years. I came along this morning out of a sense of duty. To my amazement I find I now have a new perception of ILSD.'

Outcomes

Regal Chemicals Group senior management accepted that there were new ideas and energy within its information function. They invested in it in two forms. There was capital investment in the new information technology needed to support a more proactive ILSD. Also, when the HoD who had initiated the work reached retirement age, soon after the investigation was completed, the newly appointed HoD was drawn from Regal's cadre of young up-and-coming managers. In the culture of Regal this was a significant act.

Methodologically the work illustrated four cycles of SSM use: using the simple EROS model (Figure 3.12) to settle on relevant systems to model; building models and comparing them with current reality; modelling in more detail and defining the concept of a rethought ILSD; widening the debate to seek endorsement for significant change. The first three cycles illustrate SSM (c), dealing with the *content* of the situation. The fourth – widening the debate – is a move to SSM (p), with a focus on the *process* of doing the work, necessary here because the ILSD team did not have the power to make the new concept a reality unless they had the approval of senior management. This move to SSM (p) was carried out mentally rather than formally. There was no need to make a model covering report writing, presenting to ILSD and the Regal managers, etc., but the team were at that point well aware of the changed focus of the thinking.

This work also illustrated the value of *consciously organizing the thinking process*. It was the guidelines provided by the methodology which enabled the ILSD team to move confidently, first from the EROS model to 'what' models, then to 'how' models and finally to doing something they had never done before, namely make a presentation to senior management. The trick in using SSM is to make sure that the methodology guidelines are immediately helpful *without becoming prescriptive*. If they do become a formula-to-be-followed, interest quickly wanes. This is a reminder that a methodology, properly applied, can help the users to get the most out of their thinking. It cannot, of course, *do* the thinking or 'solve the problem' for the users.

A full account of this investigation is given in Chapter 5 of the first edition of *Rational Analysis for a Problematic World*, edited by J. Rosenhead (John Wiley & Sons, Ltd, 1989). (The second edition contains a different case.) It is also the subject of 'Achieving desirable and feasible change: an application of SSM', a paper by Peter Checkland in the *Journal of the Operational Research Society*, Vol. 36(9), 821–831, 1985.

Case 4: Tackling Your Own Issues

The Situation

In the early days of SSM's development, the Lancaster researchers had no misgivings about carrying out a study of someone else's situation and making recommendations for action. Soon, as the true nature of SSM emerged, effort shifted to getting people in the situation to carry out the investigation themselves, with facilitating help, as in Cases 1 and 3 above. It is also possible in principle, of course, for SSM practitioners to tackle problematical situations which they themselves face. Case 4 is an example of that, the practitioners being the researchers developing SSM, the situation being the development of SSM itself. This was certainly 'problematical'.

The situation, in brief, was this. Gwilym Jenkins set up the postgraduate Department of Systems Engineering at Lancaster University in the mid-1960s. Initial expertise came from a statistician, a chemical engineer and a control engineer, but Jenkins interpreted the word 'engineering' liberally. In English you can 'engineer' a chemical plant, but you can also 'engineer' the release of hostages. When Peter Checkland was recruited, Jenkins asked him to lead research to answer the question: 'Can classic systems engineering, which works well in technical problem situations, be used to tackle *management* issues and problems?' (The answer, eventually, was 'No', and SE had to be transformed into SSM in order to deal with the complexity of human situations.)

The preferred way of researching to answer the question which Jenkins had posed was to work in organizations outside the university to help them tackle their current problematical situations. To make sure these research situations were truly 'real', the department formed the consultancy company ISCOL Ltd. This was wholly owned by the university, charged fees for its consultancy work, and gave the surplus it earned each year to a charity, namely Lancaster University (thus avoiding corporation tax)!

This 'action research' approach through a consultancy was possible because the one-year Masters Course at Lancaster University attracted students of average age around 30. This meant that they had significant experience of coping with organizational life. They could spend the five-month project which formed part of the Lancaster course working in a real situation in which a staff supervisor (but not the student) had a consultant's commitment to a client.

This was a very exciting research situation, but also a complex one full of tensions. The members of staff, were, when wearing one hat, working in a university department: teaching, researching, gaining new knowledge and giving it away in the open academic literature. Wearing a different hat, they were also part-time consultants working for paying clients as members of an organization which had to earn its keep in a market and also, where appropriate, maintain client confidentiality. In addition to departmental staff and mature Masters students there were also several full-time ISCOL consultants – ex-students staying on for a year or two to gain further experience. Their salaries had to be generated through ISCOL activity. There was no question of ISCOL Ltd being subsidized. The university tolerated rather than supported it. So the company had to make a surplus each year, otherwise its owner would close it down. As SSM took shape within the research programme, it seemed a good idea to step back and use it to examine the complicated situation in which the approach itself was being created.

The Use of SSM

Experienced users of SSM become well aware that relating the guidelines of the methodology to a worrying situation is a good way of distancing yourself from it mentally. This enables the situation to be examined more dispassionately, which is normally more rewarding than relying on the more usual mixture of rambling discussions, casual conversations and gusts of emotion. Also, the explicit use of methodological guidelines makes it more difficult for unstated personal agendas to dominate. This was by no means fully realized at the time of the work now being described. However, simply because that work involved some navel-gazing, it was decided to examine it using SSM in an open, explicit, flat-footed way.

The structure of the situation in which SSM was being developed is shown in the Rich Picture in Figure 3.14. Here we have two sets of activities that are carried out by two linked organizations which embody very different

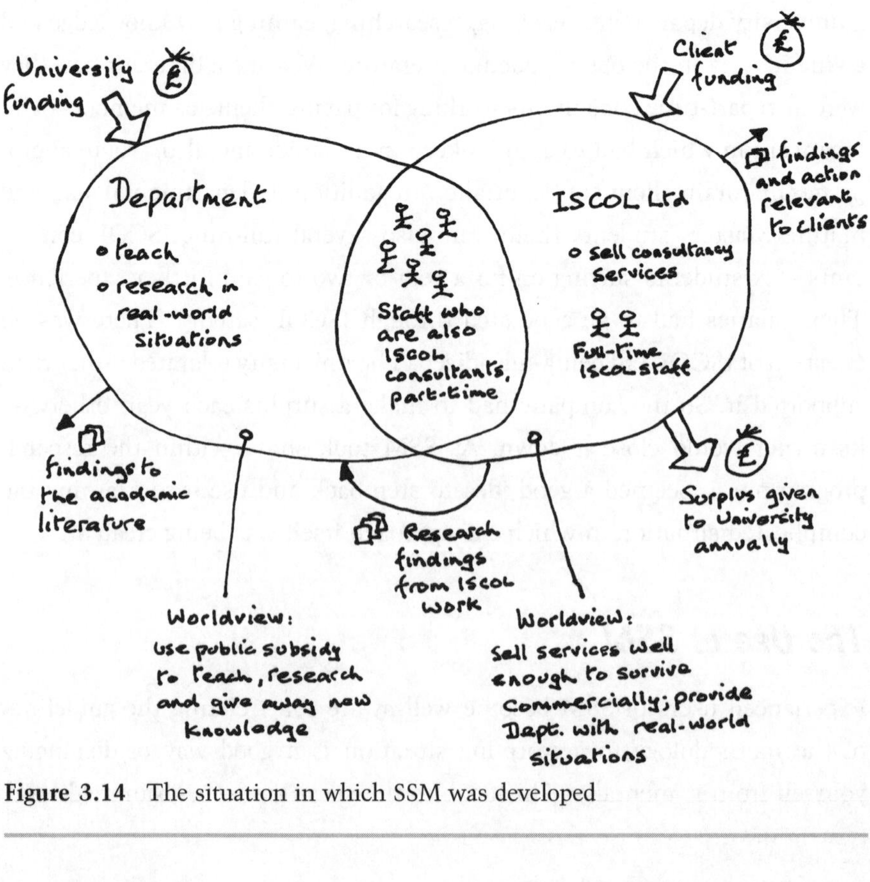

Figure 3.14 The situation in which SSM was developed

worldviews. The university department, subsidized by public money, teaches, researches and gives away new knowledge gained from its research in the open academic literature. ISCOL Ltd, unsubsidized, provides consultancy services to clients in return for fees, and seeks to survive from year to year by at least breaking even financially. Links between the two organizations are very strong. All members of staff in the department also act as part-time consultants for ISCOL (and, in addition, adhered to a 'gentlemen's agreement', as a group, that they would do no consultancy work as individuals other than through the university-owned ISCOL). Also, findings from ISCOL's experiences with clients were required to feed into the department's research agenda.

In discussion, university staff and ISCOL consultants accepted the concept in Figure 3.14 and agreed that it revealed inevitable tensions in the situation as a whole. Roles, norms and values (in SSM's Analysis Two) which are appropriate for a university department are clearly different from those of a consultancy living in a marketplace. In addition, the concept also usefully made clear the fact that in terms of power (SSM's Analysis Three), the department was dominant. The acceptance that ISCOL was not trying to maximize its income, and did not aspire to grow (as would a normal wholly commercial consultancy) indicated that ISCOL was only a means by which the department was seeking to achieve its ends – namely to engage with serious real-life situations in an action research mode. ISCOL was a 'how' related to the department as a 'what'.

Given this very close relationship it was agreed to treat Figure 3.14 as an adequate account of the situation and derive from it a model with two linked sub-systems. These contained activities relevant to the concepts 'department' and 'consultancy'. Giving each sub-system its own monitoring and control sub-system naturally revealed very different criteria for efficacy, efficiency and effectiveness (E_1, E_2, E_3). (For example, effectiveness would be measured by scholarly criteria for activities undertaken by a university department, and by client satisfaction – combined with survival – for a consultancy.)

The difficulty of naming E_1, E_2, E_3 for the model *as a whole* was a clear indication of the existence of tensions in the situation. Similarly, analysis of who in the real world would in principle carry out the activities in the model starkly revealed why the department staff felt hard-pressed. In the arrangements created, staff undertaking project supervision of students working on ISCOL projects were in four very different roles simultaneously. They were still *teachers*, now out of the classroom, since most students found the five-month project a rich learning experience; they were *assessors*, since ultimately they would be awarding marks for project performance; they were *consultants*, with a commitment to a paying client; and they were *researchers* using the project as a vehicle for research. No wonder staff felt stretched!

The structured discussion based on the model of linked 'department' and 'consultancy' activity also highlighted two other features of the situation. The extent to which the annual round of five-month projects for Masters students dominated department and – to a lesser extent – ISCOL thinking, was very apparent. Also there was discussion about increasing the link between department staff and the several ex-students who were ISCOL full-time consultants.

Out of this investigation a package of changes to improve the overall situation emerged. This happened without undue pain, and changes emerged which both the department and ISCOL could live with.

Outcomes

The most significant change to come from this investigation addressed the problem of staff workload. By deciding to assign two postgraduates to each of the Masters projects, the project load was immediately reduced. This change quickly brought with it two other benefits which had not been anticipated. First, the pair of students on a project in a client organization naturally maintained a continuous dialogue about the work over the project period. This contributed strongly to their education and understanding. Also, since students were required to write an individual dissertation, the structure of these had to be rethought. In a new format, Part One asked for an account of the work done, written for a notional audience with no prior knowledge of either the situation or the work done – which was the position of an external examiner evaluating the dissertation. Part Two then consisted of reflections on the project in which students were encouraged to relate their own slice of real-world experience to the classroom material from the two teaching terms. Essentially the student was answering the question, 'What have I personally learnt from what is intended to be a learning experience?' This provided a level playing field for all students, irrespective of whether they had found themselves in a difficult or a supportive project situation. And it delivered a flow of dissertations which not only helped with the development of SSM but also enabled classroom teaching to be steadily modified as the body of real-world experience accumulated. (A student remarked: 'It would be a good idea to retake this course every five years!')

The second change agreed was to broaden the department's research base by increasing the number of doctoral students; the third was to agree to try to strengthen the department–ISCOL link by providing 'thinking time' for the several full-time ISCOL consultants in between projects. This was relevant because (as in Figure 3.14) ISCOL was not there to do consultancy for its own sake, and the company always emphasized the university connection in its marketing. It was there to provide access to serious situations in outside organizations, something well beyond the common – more threadbare – situation in which academics visit organizations in order to 'collect data' which they then take away to analyse.

Finally this investigation led to an agreement to examine the implications of ISCOL becoming a research institute rather than a fee-earning consultancy. This was done but the idea was rejected; real *engagement* with management issues was too valuable to be given up.

Methodologically, the structured framework of SSM had the effect of engendering a calmer, more reasoned approach to department/ISCOL issues – about which individuals felt strongly – than would have occurred in a series of ordinary management meetings. That is the virtue of a methodological framework, as long as it is not allowed to become a prescription.

As a postscript to this account, we may record that although the University tolerated, rather than encouraged ISCOL, the consultancy's annual surpluses contributed nearly £0.5 m to the university coffers over the period of its existence.

This investigation is described in Chapter 7 (pp. 206–208) of *Systems Thinking, Systems Practice* by Peter Checkland (John Wiley & Sons, Ltd, 1981, 1999).

Case 5: Clarifying a Concept

The Situation

The authors of this book are naturally keen to see SSM as a taken-as-given approach relevant to any human situation which entails acting purposefully, with intention. But it is perfectly obvious that parts of SSM can themselves be useful, even if the full methodology is not used. For example, any manager, given the task of improving some purposeful action undertaken regularly by his or her organization, could usefully form some concepts of that action, looked at from the perspective of several different worldviews, carry out a CATWOE/E_1, E_2, E_3 analysis for each concept, and set these analyses against the current real-world version of the action. That would help to define possible ideas for 'improvement', even if no models were built and the rest of SSM was ignored. The case history which follows is an example of partial use of SSM. It entailed creating an activity model to define a concept at the heart of an initiative by the British Government's Department of Trade and Industry (DTI).

In the 1970s some government funding was provided for tackling problems which occurred widely across many industries and were, on the whole, neglected. Thus metal machinery in any industry wears out because of friction, and also because of corrosion, if steps are not taken to prevent such deterioration and consequent loss of efficiency. DTI committees on 'tribology' and 'corrosion technology' were established to bring about improvement in these areas. Then a report commissioned from management consultants highlighted another source of inefficiency in British industry, namely the failure to adequately maintain equipment and other physical assets. The report suggested that £500 m could be saved by paying even modest attention to that industrial Cinderella: the maintenance function.

Egged on by their success in establishing the word 'tribology', DTI made up a word for this new area of neglect. The word 'terotechnology' (from the Greek verb *térein*, 'to care for') was coined. A 'Committee for Terotechnology' was established, and the Committee established a panel under Mr Darnell, then Engineering Director of British Steel Corporation, to propose an argued definition which would indicate what was within the concept and what was not. Darnell wisely decided that general discussion among a disparate group of people was not a good way of agreeing a definition! He commissioned Checkland and ISCOL Ltd to undertake 'a systems analysis of the concept terotechnology', Checkland being a member of the terotechnology committee and of the panel. ISCOL assigned Checkland and Chris Pogson, an ex-student then an ISCOL full-time consultant, to the task.

The Use of SSM

Very fundamental thinking about the idea of terotechnology quickly reveals that it is fundamentally different in one important respect from tribology and corrosion. Both of those have a core of established scientific knowledge, considered use of which can in principle improve industrial processes. *Caring for* physical assets entails human activity. There is no base in physical science. It is a matter of deciding what coherent linked set of activities could constitute terotechnology, and what principle would unite these activities in particular, excluding others. Given this, the idea was there from the start, for Checkland and Pogson, of building a purposeful activity model which would be the core of the definition of terotechnology.

Very basic systems thinking indicates that terotechnology is one topic within wider industrial activity. The idea of equipment maintenance links to the idea of design, which might reduce or eliminate it; this brings in costs, and the balance of capital and running/maintenance costs over the lifetime of the equipment; this will link to the expected market for the product . . . etc. What boundary, therefore, could make terotechnology a coherent concept?

The work which Checkland and Pogson undertook was to interview members of the Committee and the panel, people who had been involved in DTI discussions which led to forming the Committee, and also experienced people in industry. The idea was to find out what concepts were evoked for them by phrases like 'maintenance function', 'maintaining plant and other equipment', etc. This constituted the 'real world' against which we could compare possible activity models.

The thinking leading to model building started from the notion that owning and using physical assets is ultimately concerned with either generating wealth (in an industrial company) or minimizing costs (as, for example, in a hospital). This led to a first concept in terms of 'a system to ensure that the care of physical assets makes the best possible contribution to generating wealth or minimizing costs'. Discussions with the panel focused on the idea of a system which acquires physical assets, cares for them in use, disposes of them optimally (when the cost of maintaining them becomes too great) and learns from the experience so that it and other functions (such as design) can improve in the future. A breakthrough in these discussions came when a distinction was made, for analytical purposes, between the costs of *ownership* of a physical asset and the costs of *using* the asset.

Take the case of a chemical plant. There will be costs arising from acquiring, using and disposing of it at the end of its useful life. And there will be many costs which arise from using it, including the cost of operating it and the cost of managing its operation. In order that terotechnology should focus on *caring for* a physical asset, and so avoid becoming a hazy version of the broader concept of *managing* the asset, it was decided that the difference between 'owning' and 'using' a piece of equipment should be a demarcation between what was part of the terotechnology concept and what was not. Terotechnology should refer to acquiring, caring for and disposing of physical assets. Its measure of performance would stem from the decision to dispose of and/or replace it. The 'terotechnology system' would try to minimize the total costs of *owning* a physical asset over the whole life cycle of it, and, indeed, would take the decision concerning when to get rid of, and/or replace the asset.

The final agreed model of 'the terotechnology system' is shown in Figure 3.15. In SSM language this is an issue-based rather than a primary-task model. Minimizing the total life-cycle costs of owning equipment, etc. is an important issue for management. The activities in Figure 3.15 have to be carried out in any competently managed organization, and will involve many people: managers, designers, engineers, accountants, information specialists, etc. But they are not usually thought of as a function to be institutionalized

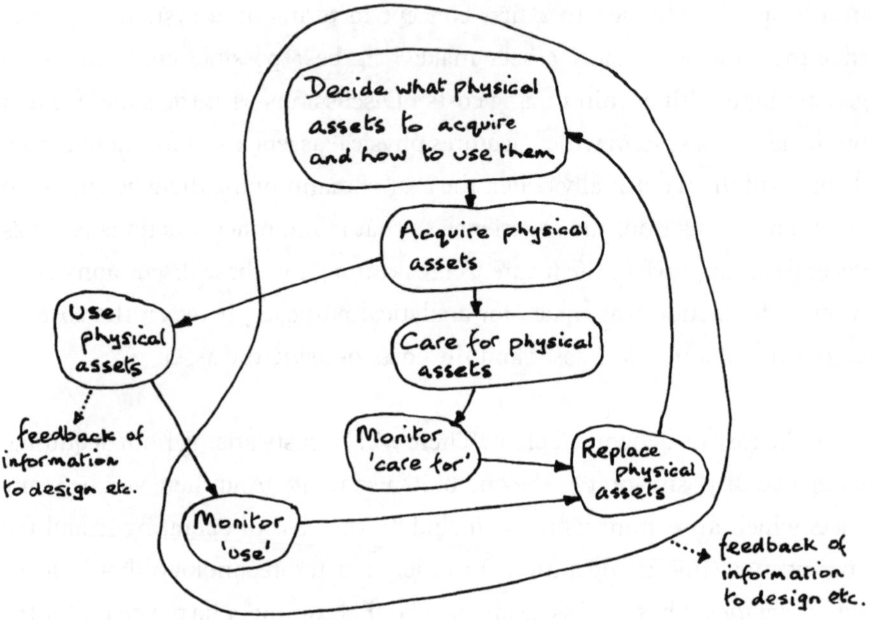

Figure 3.15 Terotechnology in outline, as a system

in a section or department. Another feature of the model is that it does not apply to a particular scale of asset-owning. It will be highly relevant, but not restricted to building and operating a shipyard or an oil refinery. In principal it applies 'to a pen as well as a printing press', as Pogson put it.

Outcomes

The terotechnology panel reported their work, culminating in the 'terotechnology system' model, to the main Committee. The Committee accepted the arguments leading to the model, and the model itself, as the basis for DTI's definition of terotechnology. The Ministry's publications in this area defined it as follows.

> Terotechnology is a combination of management, financial, engineering and other practices applied to physical assets in pursuit of economic life-cycle costs. Its practice is concerned with the specification

and design for reliability and maintainability of plant, machinery, equipment, buildings and structures, with their installation, commissioning, maintenance, modification and replacement, and with the feedback of information on design, performance and costs.

One of the current short Oxford dictionaries defines it as: 'the branch of technology and engineering concerned with the installation, maintenance, and replacement of industrial plant and equipment'.

So the word and the concept have survived. If you buy an old car extremely cheaply and then find you are spending a fortune keeping it on the road, you face a terotechnological dilemma. If you open the bonnet and find that the nut you need to unscrew is unreachable by any known form of spanner, you know that the motor manufacturer has neglected terotechnological considerations in design.

Methodologically this work simply used activity modelling as a way of making sure that discussion of an abstract concept was coherent. The initial model was used to explore material from interviewees, and refined until the members of the panel were satisfied that a final version was both well-founded and defensible.

An account of this investigation is to be found in Chapter 7 (pp. 202–206) of *Systems Thinking, Systems Practice*, Peter Checkland (John Wiley & Sons, Ltd, 1981, 1999).

4
SSM in Action in the Field of Information Systems

Introduction

It was obvious throughout the development of SSM that in most management situations issues related to information provision were never far away. It was also obvious that SSM had something to contribute in this area. So it is not surprising that SSM has been much used in work concerned with information systems (IS) and information technology (IT). The relevance of SSM here comes from the fact that as soon as you have an SSM-style activity model you can move on to ask of each activity in it: What information would in principle have to be available to someone carrying out this activity? Also: What information would be generated by doing this activity? (Then follow many subsidiary questions of the kind: Who would supply or use this information? In what form? with what frequency? etc.) Thus an activity model can be transformed into a related information model, and the questions asked of the real world will be information-oriented rather than or in addition to activity-oriented questions.

This chapter describes examples of SSM use in the field of information systems, but before moving on to those accounts it is appropriate to describe the view of the information field which 'SSM thinking' supports. That perspective underlies each of the case histories included here. Two key aspects of it will be drastically summarized below, as a basic requirement for appreciating the cases. The argument is set out in detail in *Information, Systems and Information Systems* by Peter Checkland and Sue Holwell (John Wiley & Sons, Ltd, 1998).

Data, Information, etc.

There are uncountable millions of facts about the world – items of *data*. Pluck out a random example: the temperature at the airport in Reykjavik this afternoon, as this sentence is being written. This temperature is undoubtedly recorded, and no doubt could be quickly discovered given a few minutes on the Internet. We may have no interest in doing this, but if we were flying to Iceland tomorrow and were about to pack our clothes for the visit, we might well look up the temperature in Reykjavik. The fact would have acquired a new status: no longer simply 'an item of data' which we expect to exist in principle, but 'an item of data *which we have an interest in knowing*'. It is to be regretted that we do not have a word which marks the important distinction just made. We normally use the word 'data' to cover both categories, alas. Checkland and Holwell suggest the word 'capta' for that subset of all data which we have an interest in knowing ('capta' being made up from the Latin *capere*, to take, just as 'data' comes from the Latin 'dare', to give). Having paid attention to some data, thus turning it into capta, we enrich it further as a result of our interest in it. That provides a context which makes the capta *meaningful* for us. We call such meaningful capta *information*. Finally we can assemble items of information into larger, multi-layered, longer-living structures of *knowledge*. Figure 4.1 shows these relations between data, capta, information and knowledge.

Even this simple bit of thinking illustrates important points. For example, in the field of information, as computers became common, the reasonably accurate phrase 'data processing system' was always used (though

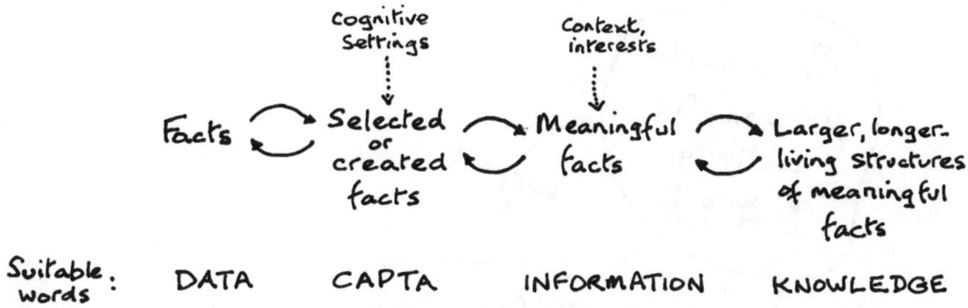

Figure 4.1 The relations between data, capta, information and knowledge

'capta processing system' would have been more accurate). Unfortunately the phrase evolved into 'information system', or 'management information system', no doubt because that has a more up-market ring to it. But 'information' is created when *human beings* attribute meaning to data in a particular context. No machine operations can do that. A system designer will of course have some defined information provision in mind; but the designer cannot determine how the system will be used. Thus, consider a spreadsheet of student marks on a course taught by several teachers over a university term. This ostensibly provides information concerning student assessment. But lecturers may well use it in internal battles over uneven teaching loads! Machines can be programmed to manipulate electronic tokens for data, but it is human beings who confer meaning.

Information Systems

No one creates an information system for its own sake. The nature of an information system is that it helps or supports or enables someone carrying out some purposeful activity. Now, when one system, A, supports another, B, it is obvious that you cannot conceptualize or design A until you have first conceptualized B. Thus if the managing director of a company asked you to design a system to do research to support the company's operations, you could not make a move until you had a clear idea of what the company did. When you knew that you could conceptualize what 'research support' would mean in that company; only then could you begin to think about the 'system to do

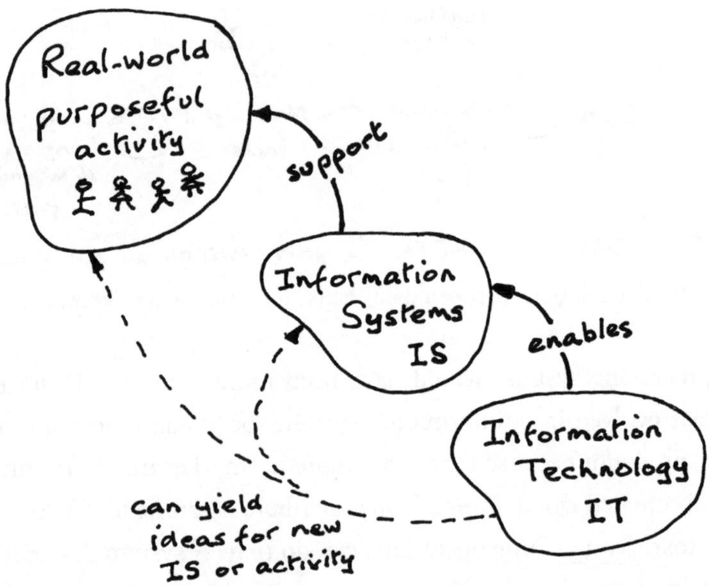

Figure 4.2 Relations between IT, IS and real-world purposeful activity

research'. This is very basic systems thinking, though it is often neglected. But the message is clear: in order to do work on information systems you must first focus on defining carefully the purposeful activity which the information system will support. The general situation is illustrated in Figure 4.2.

Nowadays most information systems will be realized through computer-based technology. That enables the information system to be realized; and it supports some purposeful activity in the world. Obviously sensible thinking must start with careful definition of the purposeful activity served. Especially to be avoided is the procurement of a complex IT system without first thinking out exactly how it will be used, as well as what activity it will serve, though every reader will know of organizations in which that has happened. Are there any in which it has not? But it must also be noted that the nature of modern IT is that it may make possible purposeful activity and information systems which would be impossible without it, hence the dotted arrows in Figure 4.2.

That completes the necessary preliminaries. Case histories of work in the information systems field using SSM now follow.

Case 1: Creating an Information Strategy in an Acute Hospital

The Situation

In the early 1990s the Royal Victoria Infirmary in Newcastle, a large acute hospital, and Hexham General, a much smaller hospital 25 miles away, merged to form a single organization, while remaining on their separate sites. An innovative Information Officer of the proposed new hospital, Steve Clarke, seized the opportunity provided by the merger to reformulate an information and IT strategy for the combined hospital. He obtained some central National Health Service (NHS) funding for a project to do this, put the work out to tender, and awarded the contract to a large management consultancy. They assigned John Poulter, one of their senior consultants, to the project. Clarke wanted an SSM-based participative study carried out by professionals from the Hexham and Newcastle hospitals with facilitating help. This help was provided by Clarke, Poulter and Checkland, the latter being brought in to provide methodological guidance. Checkland saw this as a good opportunity to help with the use of SSM by busy hospital professionals, most of whom had never done anything like this and who had never heard of SSM. He was somewhat daunted to find that, with work starting in March, the new information strategy was to be presented to the Board of the hospital in September. This gave an edge to the project; there would be no time for speculative diversions.

The Use of SSM

This work was carried out at a time in the NHS in which 'purchasers' of health services for a given population (they were previously local

Health Authorities) made annual (!) 'contracts' with 'providers' such as the Newcastle hospital. This was part of a short-lived attempt to create an 'internal market' within the NHS via these 'contracts'. They were ill-named, since they were not legally binding, and were soon replaced by service agreements over several years. (The purchaser/provider split has remained.) At the time of this project Newcastle had to find purchasers to cover its costs of more than £70m p.a.

The project was initiated at a meeting of a hundred staff addressed by the Chief Executive and the facilitators. The Chief Executive emphasized that the new information strategy would emerge not from 'management' but from the work of small groups of hospital professionals (about 40 in all) drawn from all areas of hospital activity: doctors, nurses, people from the estates office, accounting, etc. The facilitators then explained how the work would be done. Each group would be mixed in composition, and would be asked to examine the core purpose, activities and information needs of several of the functions performed at the hospital's two sites. There would be a 'plenary meeting' each month at which representatives of the groups would describe progress and problems for discussion with the facilitators, who would also attend some meetings of the groups. The whole exercise would be kept under review by a steering group of directors and senior managers from the two sites, and would be completed by September.

This plan reflected specific features of the situation: a tight timescale and a specific outcome, namely a reformulated information strategy based upon its current pattern of activity, rather than a broad investigation of the strategic future of the hospital. Methodologically the greater difficulty was the ignorance of SSM among the doctors, nurses, etc. in the small groups. On the other hand, given the required outcome, this investigation would focus on primary-task models rather than issue-based ones. It was therefore decided that Checkland would make some *generic* activity models relevant to being an acute hospital. These would be offered to the groups for them to use as a starting point. They could adopt them or modify them as they thought fit, based on their experience. The only aspects of SSM presented to the group members, apart from its general role in information-focused work, were CATWOE analysis and the measures of performance for T, namely the criteria to judge efficacy, efficiency and effectiveness (E_1, E_2, E_3).

A generic model, in principle, relevant to any acute hospital, was made. Its Root Definition, CATWOE and E_1, E_2, E_3 were:

> A system, operating under a range of external influences, which, in the light of a strategy based on its capabilities and costs, delivers services defined in 'contracts' with purchasers within the context of NHS norms and policies, that service delivery itself contributing to the ongoing development of its strategy for service provision.

C	those receiving hospital services; purchasers
A	hospital professionals
T	need for acute services → need for acute services met
W	acute services can best be provided by an organization dedicated to developing and delivering such services
O	hospital management board; NHS executive
E	NHS structures and norms; the purchaser-provider split
E_1, efficacy	demonstrable delivery of a portfolio of services of suitable quality
E_2, efficiency	minimum use of resources (expressible in money and time)
E_3, effectiveness	satisfaction of patients treated, purchasers, the NHS executive; contributions to hospital reputation (i.e. contributions to long-term viability).

The model itself is shown in Figure 4.3. In that model, four of the five main activities (the 'magical number' 7 minus 2 in this case!) are expanded at a

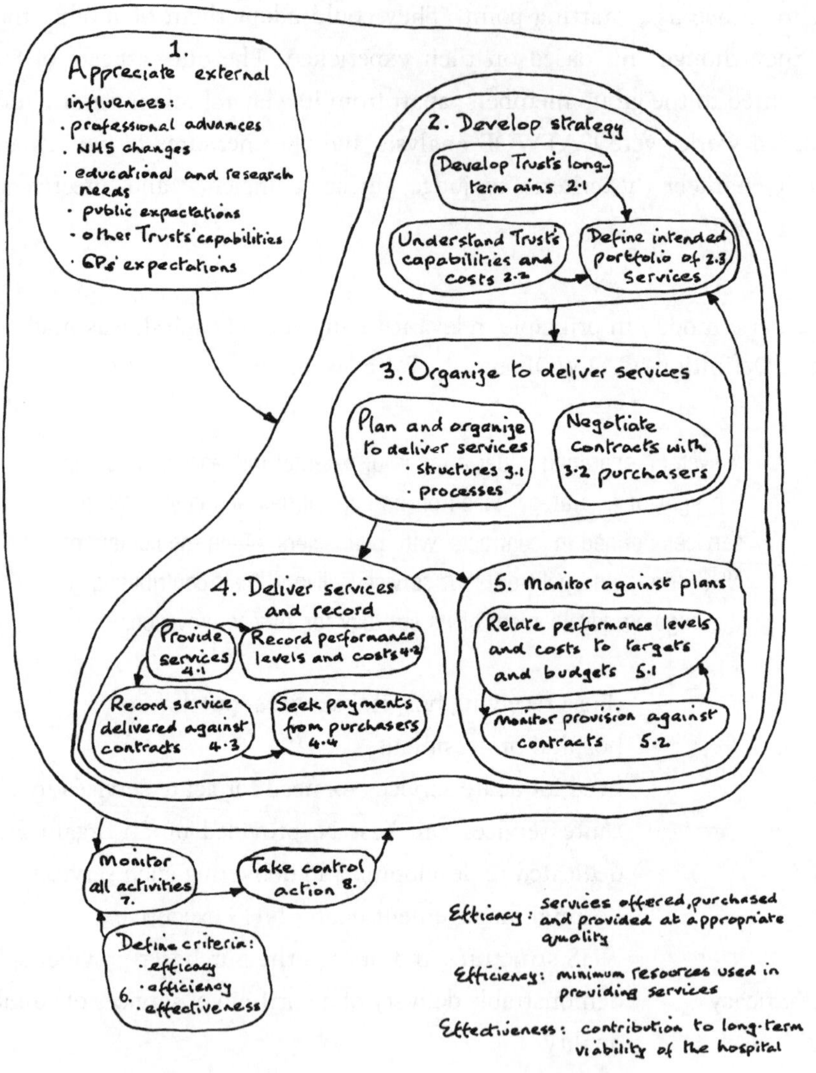

Figure 4.3 A generic model of the concept: 'an acute hospital'

second level. Carrying out the expanded activities 2, 3 and 4, and monitoring that activity, would yield information about the hospital's capabilities and costs. This, together with an understanding of the external influences in activity 1, would provide a contribution to the development of hospital strategy (activity 2); hence the internal feedback from activity 5 to activity 2.

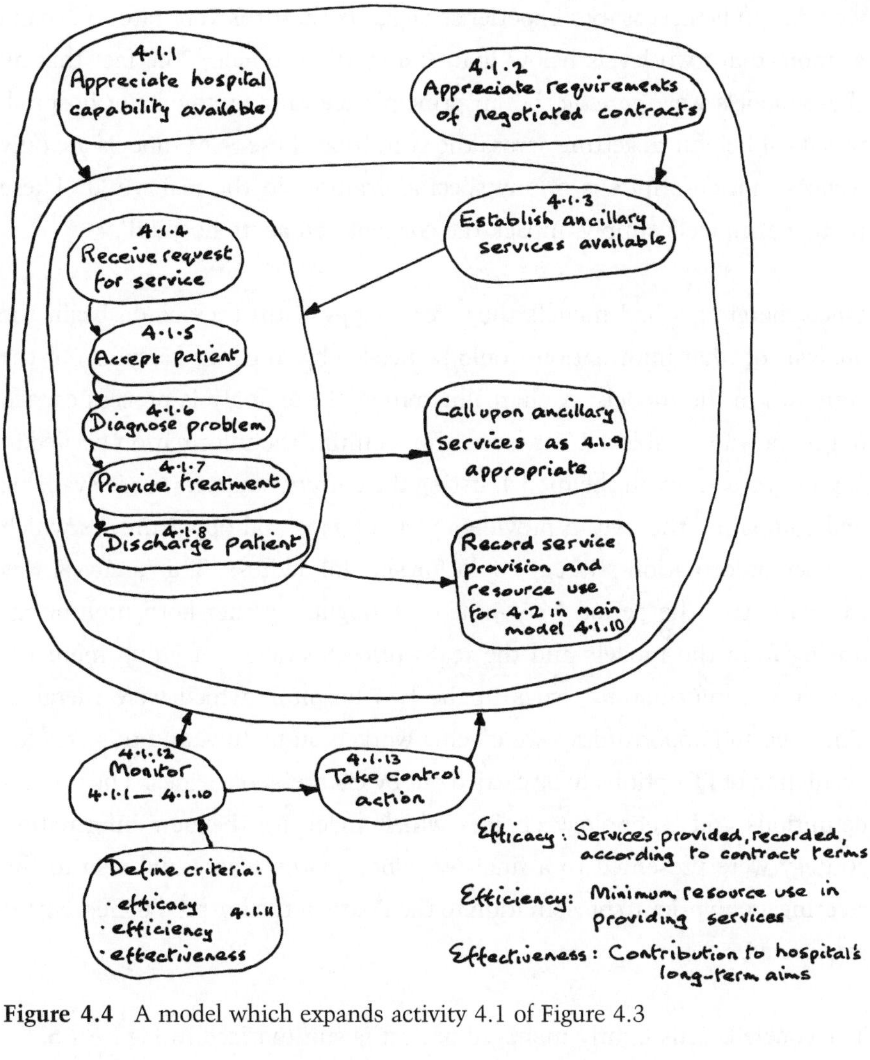

Figure 4.4 A model which expands activity 4.1 of Figure 4.3

This is still a high-level model, so further models expanded activities 2.1, etc. in more detail.

For example Figure 4.4 shows further expansion of activity 4.1. A few activities from the expanded models like that in Figure 4.4 were themselves expanded at a fourth level of detail, but no complete models at that level were felt to be necessary. (Experience suggests that it is very rare to expand at more than two levels below that of a starting model.) The fact that all these models were *generic*, i.e. in principle relevant to any acute hospital, was itself helpful in getting across the point that these SSM models are only devices, not descriptions of any specific example in the real world. These particular models simply unpack the concept: 'an acute hospital'.

Once the groups had models they were happy with, they could begin the analysis of what information would be needed by someone carrying out the activities in the models. A chart illustrative of the analysis process carried out is shown in Table 4.1. Its focus is on defining the information needed to support an activity in the model, listing the current support that is available and comparing the two to provide a view of gaps and opportunities which the new information strategy could (or should) address. This analysis was carried out in the period May–June. It brought together both promptings arising from the models and the real-world experience of group members concerning information issues in the two hospitals which were merging. The gaps and opportunities were being worked on by July, and this enabled evaluation of IT options to be carried out by Clarke's colleagues. The process definitions and technology choices which made up the new information strategy were presented to a final workshop of the teams and also to the steering group prior to presentation to the Board of the hospital in September as planned.

The course of this tightly-managed project is summarized in Figure 4.5.

Outcomes

The practical outcome was of course the delivery of the new information strategy on time. Methodologically this investigation was relatively straightforward, given that the timescale dictated that it would be based on existing patterns of hospital activity questioned via primary-task models. In the time

Table 4.1 An illustration of the kind of chart used for information analysis

Activities from the model	How the activity is done	Measures of performance	Information needed	Information support provided by	Information gaps and opportunities
4.1.4 and 4.1.5 Receive request for service, and accept patient	Letter, phone call	Speed with which the request is handled	Patient's details, clinical condition, and history Contract situation	Patient administration system (PAS)	Automatic generation of letters to patient and referrer Up-to-date contract situation
4.1.6 Diagnose problem	Consider history Examine patient Conduct investigations	Medical audit	Case notes Results from investigations		Case notes often missing Much duplication of recording of patient's details Delays in receiving test results
4.1.7 Treat patient	Conduct procedures/operations Prescribe drugs	Medical audit	Availability of facilities (theatres, anaesthetists, etc.) Drug effects and interactions	Theatre booking system	Systems not available at ward level
4.1.8 Discharge patient	Discharge summary Discharge letter	Speed with which produced	Post-treatment test results Availability of discharge facilities Coding	PAS	Links to ongoing providers of care Automatic generation of discharge summaries and letters Support for Read coding

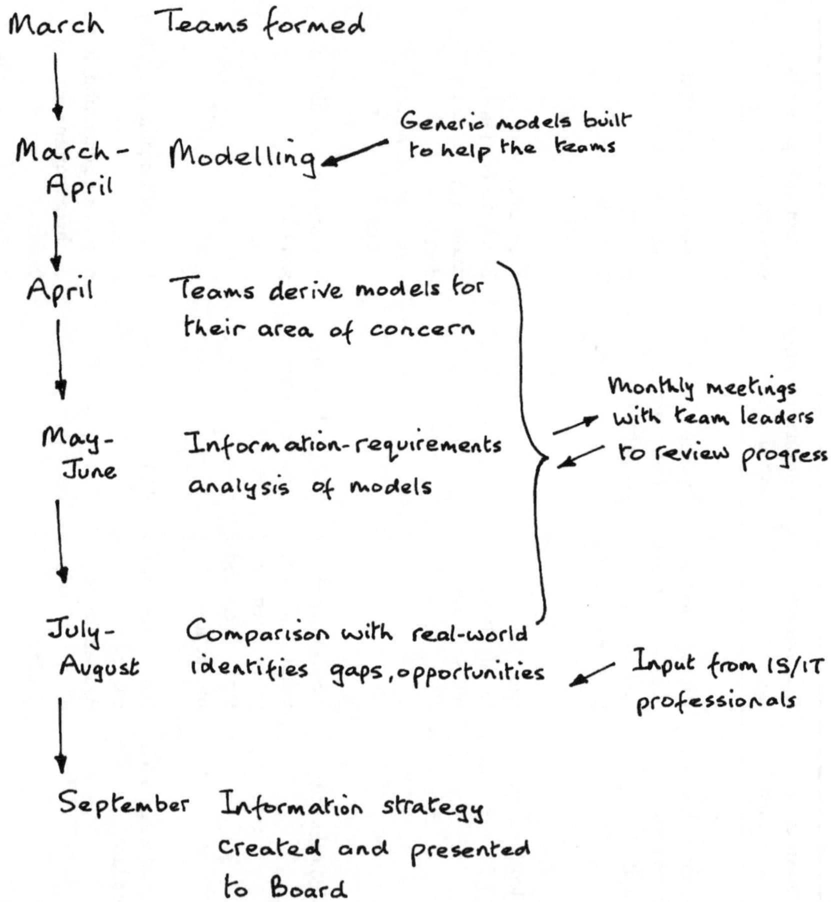

Figure 4.5 The course of the project to create a new information strategy for the hospital

available it could not have become a more luxurious study of future hospital strategy, which would have required much more time and effort as well as more models, including ones which were issue-based. The greatest problem was meeting the deadline, and that was achieved by the careful management of the process by Clarke and Poulter. This ensured that the momentum was kept up, even though members of the working groups were having to fit in this extra work alongside their normal work. Luckily they found it interestingly different from their day-to-day work.

Lack of knowledge of SSM was not a problem. This example of SSM use in fact illustrated the general finding that people who learn the approach by using it find SSM easy to understand in a fairly deep way.

This story is a good example of the LUMAS model in action (Figure 1.8). Steve Clarke's understanding that SSM is best used participatively led to the adaptation of the methodology to the particular approach used here, namely small-group working with primary-task models at several levels of detail followed by information analysis.

Two specific happenings illustrated how the organized use of a declared methodology can broaden thinking and lead to learning. At one of the monthly plenary meetings a senior nurse, whose group were examining what is entailed in providing a nursing service in an acute hospital, was anxious to tell the rest of us what happened at one of her group meetings. They decided to define CATWOE for 'a system to provide a hospital nursing service', prior to thinking about modifying one of the generic models which had been provided. Answering the 'C' question (Who would be beneficiaries or victims of this purposeful system?) they naturally at once said 'patients', and moved on. They were pulled back by a member of the group who suggested that under the 'internal market' arrangements – annual contracts with a purchaser – you could argue that the C of CATWOE was not 'patients' but the hospital 'contracts manager', with the nurses' responsibility being to supply the amount and quality of nursing care that was paid for under the contract! This led to long and animated discussion. The senior nurse who told this tale said that she had known that she was against the contracting regime, and had had the feeling that the professional autonomy of her profession was being eroded, without being able to pinpoint her concern. This CATWOE discussion explained to her why she felt as she did.

The second example of surprised learning came from a paediatrician who said, 'I've worked here for 20 years, and this is the first time I've ever had any sense of the total activity of the hospital as a whole.' Both these examples illustrate that although no methodology can itself produce learning and 'answers', as if by magic, organized use of methodological guidelines can render the users' own thinking more coherent and more effective than that which occurs in the day-to-day hurly-burly of organizational life.

More detailed accounts of this work are to be found in Chapter 7 (pp. 206–213) of *Information, Systems and Information Systems*, Peter Checkland and Sue Holwell (John Wiley & Sons, Ltd, 1998); in Chapter 5 (pp. 91–113) of *Rational Analysis for a Problematical World Revisited*, J. Rosenhead and J. Mingers (eds) (John Wiley & Sons, Ltd, 2001; i.e. the second edition of this book); and in Checkland, P., Clarke, S. and Poulter, J. (1996) The Use of SSM for developing HISS and IM & T strategies in NHS trusts, in B. Richards and H. de Glanville (eds), *Healthcare Computing* (BJHC Ltd, Weybridge).

Case 2: Evaluating a Clinical Information System

Introduction

This case history and the one which follows both come from uses of SSM in the evaluation of complex projects 'after the event' in order to make judgements about them. Evaluation has in recent years emerged as a significant topic in the management field, and since the phrase 'a project' conveys the idea of achieving some declared aim through organized purposeful activity it is not surprising that SSM can be used in the *post hoc* evaluation of significant projects. Both cases concern what was known as the 'Resource Management (RM) Initiative' in the National Health Service (NHS), a project on which £300m of public money was spent. It was concerned with both information systems and related organizational change within hospitals. This case summarizes the evaluation of a pioneering clinical information system in a large hospital, Huddersfield Royal Infirmary. Case 3 which follows then describes the national evaluation of the RM initiative as a whole.

The Situation

The RM initiative was launched when the Director of Finance on the NHS Management Board (later the NHS Executive) – who was on secondment to the NHS from the accountancy firm Price-Waterhouse – discovered how little hospitals knew about their resource use and the costs of various medical treatments. The Health Notice which launched the RM initiative was entitled 'Resource Management (Management Budgeting) in Health Authorities' but the phrase 'Management Budgeting' was quickly dropped. At that time the attitude of clinicians towards the concepts 'management' and 'budgeting' was: 'nothing to do with us'! (The RM initiative was one of the developments which helped to change that culture.)

Nationally the RM project provided funds for two linked developments: the creation of computer-based systems which would make knowledge of resource use and costs in hospitals readily available, together with the work on organization development which would embed this knowledge in the hospital's management processes.

At Huddersfield Health Authority, under the inspirational leadership of their chairman, the late Peter Wood, the thinking was ahead of that of the NHS Executive. When RM was launched, work had already been underway in Huddersfield Royal Infirmary for several years on a 'clinical information system' (CIS) which would bring together managers and clinicians. (It was retrospectively given the status of one of the six national pilot projects in the RM programme.) With a sophisticated appreciation of attitudes at the time, Peter Wood had boldly decided that their CIS project would be initiated without any tight definition of an ultimate aim. The project would learn as it went along. It would proceed at the pace and in the direction that the four hospital consultants involved were prepared to go along with. This was a calm recognition of the fact that any project which lacked consultant support would be dead in the water. Eventually a startling – and very much a pioneering – CIS began to emerge from a rather confused and not-well-documented experience: no formal minutes of meetings, no action plans, definitely no critical path diagrams! At that stage Wood asked ISCOL Ltd (the consultancy company of Lancaster University's postgraduate Department of Systems) to make a study of the fledgling CIS and spell out its structures and processes.

The Use of SSM

This work was not a conventional use of SSM – if there is any such thing. The aim was to find out, from the recollections of those involved and observation of current activities, what had gone on in the CIS project and to describe the emerging system clearly. It was thought initially that relevant models to guide and structure this investigation of the CIS development would be activity models of doctors treating patients, etc. However, it was discovered that the CIS was actually a system which *monitored* the taken-as-given 'diagnose and treat' activity, *took control action with respect to it*, and *learnt* from the experience so that performance could be improved over time. It gradually accumulated knowledge of the health gain obtained from a given use of resources, and could work to improve that. Figure 4.6 shows the form and processes of the system.

In this doctors (hospital consultants) provide an anticipated profile of care for a given condition (e.g. uterine fibroids) named using the classification system known as the 'Read codes'. (This was an early use of the Read classification system, well ahead of its later adoption throughout the NHS.) This profile covered symptoms, possible complications, treatments, expected length of hospital stay if appropriate, and expected outcomes. Managers

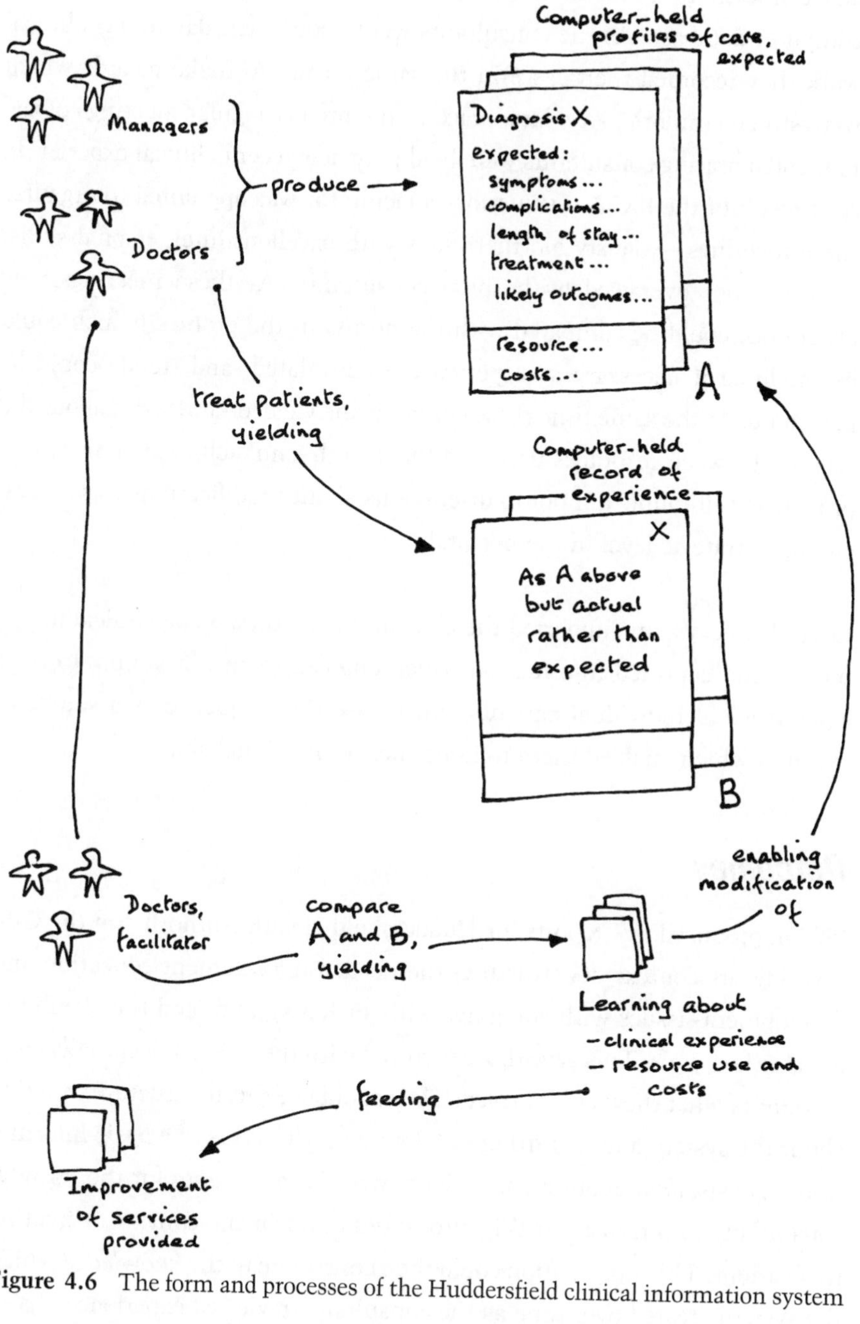

Figure 4.6 The form and processes of the Huddersfield clinical information system

added information on resource use and costs. Such profiles were held in a computer database. As the consultants went about their day-to-day clinical work they recorded their cases in the same format to make records which were stored similarly. At a later stage in the project regular meetings of the four collaborating consultants were held to review recent clinical experience. As a result of the ISCOL evaluation, a facilitator was appointed to organize these meetings – a feisty Northern lady with excellent interpersonal skills, who was not overawed by hospital consultants. At these meetings new experience could be compared with the norms in the profiles (which could be modified if necessary as experience accumulated) and trends could be picked up. At the same time these comparisons yielded knowledge about the relation between resources used and the 'health gain' achieved. This would provide a well-founded input to discussions about modifications to services, etc. at a strategic level in the hospital.

Overall this system supported the consultants as they made clinical judgements, and captured information which enabled them not simply to treat a sequence of individual patients, but to use that sequence as a source of research which enabled them to hone their professional skills.

Outcomes

ISCOL produced five reports for Huddersfield Health Authority on the CIS, covering: its context; its structure; the needed management education and development; issues with going live with such a system; and transferability to other hospitals. This gave those involved with the CIS a conceptually clear account of what they had achieved. They could use this to structure thinking about the system and its further development. (In terms of a basic information analysis, *data* about a particular patient became *capta* for the hospital consultant, who transformed it into *information* in the context of treating this patient. This information could then contribute to the *knowledge* which the system created over time as the consultants reviewed experience.)

Methodologically the most interesting learning concerned the process by which the CIS had been created. It was far from the conventions of textbook 'project management' but was absolutely appropriate in seeking change well outside the existing norms of NHS culture at that time. It reinforced the message which is built into SSM, namely that desirable changes have to be *feasible* for these people in their specific situation with its own history and sense-making narrative. This is so even though the SSM practitioner knows that what is feasible will itself be changed by the process of inquiry being used. Also, the appointment of the facilitator to organize the clinical review meetings with the consultants underlined that any good installation of an information system should include organized arrangements for making sure that learning from the operation of the system is captured and can be acted upon to improve future performance. It goes without saying that the most significant feature of the whole process of achieving the Huddersfield CIS was the ongoing joint work by doctors and information professionals to develop it. This was quite outside the threadbare norm in which system designers design and then try to impose the system on apparently awkward and obtuse users, who naturally resist.

Overall this experience left Peter Checkland and John Hardy, who carried out the evaluation, with the impression that, in spite of the absence of formal project management structures, Peter Wood, a natural systems thinker, had nevertheless guided the development of the Huddersfield CIS via an informal use of the kind of thinking which SSM brings to bear. This supports the argument that the learning approach of SSM is a 'natural' way of thinking.

A more detailed account of this work can be found in Chapter 7 (pp. 189–198) of *Information, Systems and Information Systems*, Peter Checkland and Sue Holwell (John Wiley & Sons, Ltd, 1998). The argument for the 'natural' nature of SSM is mounted in the final chapter of *SSM in Action*, Peter Checkland and Jim Scholes (John Wiley & Sons, Ltd, 1990).

Case 3: Evaluating a National Initiative in the National Health Service

Introduction

The previous case described the evaluation of a pioneering clinical information system developed at Huddersfield Royal Infirmary as part of an initiative by the local Health Authority. This local project preceded, but later became part of, the national Resource Management Initiative, which entailed more than 250 RM projects across the NHS. Money was made available centrally for projects to create information systems for monitoring the resource use and costs entailed in clinical activity in hospitals. The concept also included the required organizational development to embed the information systems in the local hospital culture. This case describes the evaluation of the national programme.

The Situation

The evaluation was commissioned by the NHS Executive following a tendering process. The contract went to a consortium led by the Health Services Management Unit (HMSU) of Manchester University. It consisted of staff from HMSU and management consultants in several fields: organization development, accounting, and computing. Peter Checkland was invited to join the group with responsibility for methodology. This account will focus on the methodology of the evaluation, which made flexible use of SSM.

The Use of SSM

The evaluation of a project that had involved more than 250 sites and the expenditure of £300m was inevitably a rather complicated business. It had to take into account activity at national, regional and site levels, and had to cover many facets of the work carried out with RM funding. These included: organization development; information systems and information technology; project management at sites and at the centre; outcomes and benefits realization; finance; and the relevance of RM to the introduction of the purchaser/provider split in the NHS, which occurred during the lifetime of the RM initiative. Not surprisingly, the 'product' from the evaluation was an HMSU report of more than 200 pages together with a second report on disseminating the results across the NHS.

Methodologically the problem was to create and maintain a coherent pattern of wide-ranging activities at several levels, the point of the pattern being to show – as the work was being done – how each different piece of work contributed to a coherent whole. SSM (p) (i.e. SSM addressing the question: How shall we go about this investigation?) and SSM (c) (How shall we create and then capture the *content* of this evaluation from the huge mass of material collected?) were equally important.

Figure 4.7 shows the result of the SSM (p) work. It takes the form of an informal model which sets out the various chunks of activity and indicates how they link and make their contribution to the whole study. Health Notice 86/34 from the Department of Health, which launched the RM Initiative, is the starting point (1). It led to three SSM-style activity models (2), which yielded questions which structured both 16 four-day visits by small teams to RM sites (3) and a questionnaire sent to more than 200 RM sites (4). Meanwhile 30 interviews with people concerned with the RM initiative at national and regional level (5) were conducted. The site visit reports produced a mass of material which was analysed using a model of 'an RM-ed site' (6). Material from (3), (4) and (5) was brought together and helped to define more detailed study, by tasked teams, of aggregated material under the headings: benefits; IS/IT; finance; organization issues; and impact on the purchaser/provider split (7). The task team reports from (7) then contributed to the substance of the main report (8).

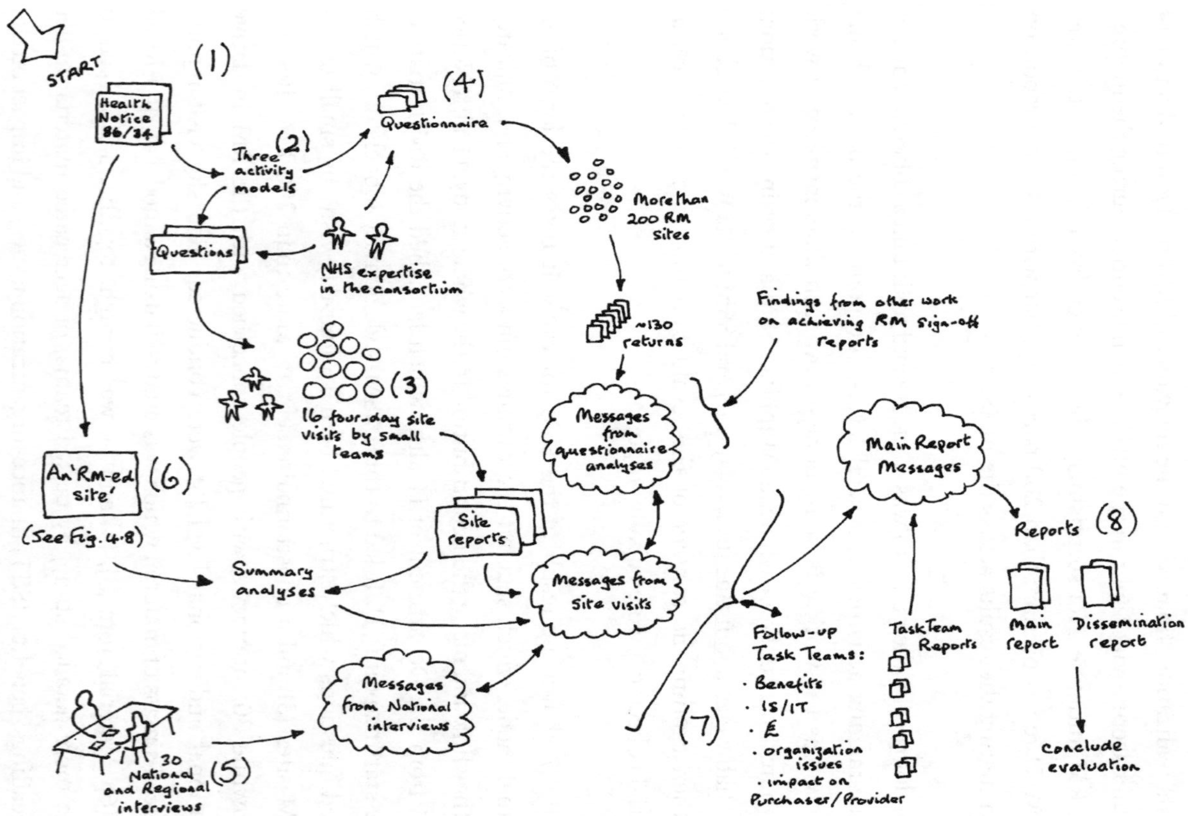

Figure 4.7 The structuring of the RM evaluation by use of SSM (p)

The work using SSM (c) was there to ensure that the mass of material which the investigation would generate would be in a form which would produce learning straightforwardly. It was based explicitly on the Health Notice which launched the RM project. The notice set out the RM concept in its 12 pages. This was taken to be an elaborate Root Definition (RD) of RM. From it, three relevant RDs were formulated and modelled. The concepts modelled (item (2) in Figure 4.7) were:

1. a system to set up and execute an RM project, including all aspects of RM as described in the Health Notice;
2. at a more detailed level, a system to *manage* an RM project; and,
3. at regional level, a system to fund and monitor RM projects, since central RM funds were channelled through the NHS regions.

These models, together with the knowledge of NHS norms acquired by the consortium partners as a result of previous work within the Health Service, were used to generate the questions which were asked and answered during the site visits (item (3)) and were posed in the 200 questionnaires, of which 130 were returned. The visits by small teams to NHS sites naturally generated a great mass of material. The problem then was to analyse this material in a way which would generate findings 'typical of RM projects' even though there were inevitably various interpretations of 'an RM project' among the many NHS sites. The solution was to go back to Health Notice 86/34 and create from it an idealized account of what the pattern of activity would be at a site which took the Health Notice as a sacred text and fully adopted all of its precepts. (In real life none did!) This model, item (6) in Figure 4.7 is shown in Figure 4.8.

It is an activity model but is expressed 'the other way round' compared with SSM's normal models. There the elements are centred on verbs, and the arrows show dependency of one activity on another. To distinguish this model from the three in item (2) of Figure 4.7, its elements are entities, while the arrows carry the activities. This model was

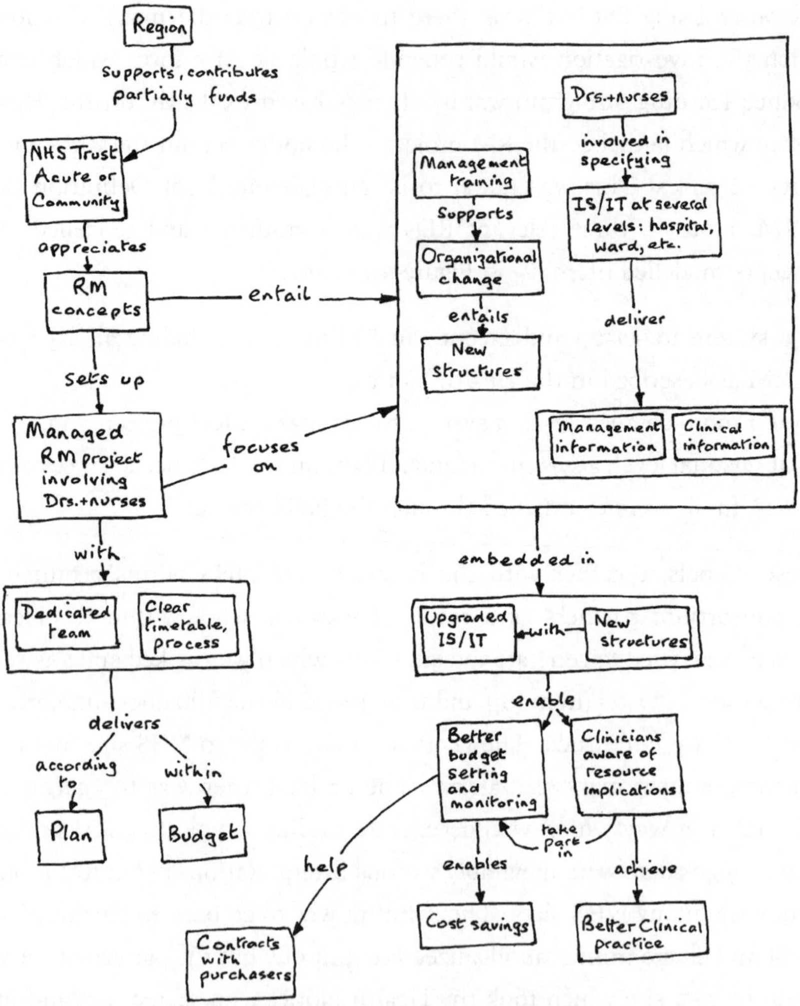

Figure 4.8 A model showing how full adoption of Resource Management would affect an NHS trust in a purchaser–provider relationship

used to interrogate, and so structure, analysis of the reports from site visits. It made inter-site comparisons straightforward, and revealed general patterns. Without it any general lessons from the site visits would have had to be obtained intuitively, simply by reading and rereading visit reports.

The reports from the evaluation consisted of the main report and another making suggestions for dissemination. Ultimately, a 40-page summary of the main report was made available to all NHS hospitals and Health Authorities.

Outcomes

At the time of the RM Initiative the NHS was undergoing noticeable cultural change. Clinicians were beginning to accept that, in an NHS with finite resources, the myth that clinical decisions could always be taken on exclusively clinical grounds, as if resources were infinite, could not be sustained. Clinicians were beginning to be comfortable with the idea that they were in fact managers of resources, and had to attend to management issues. (A few years later the setting up of the British Association of Medical Managers 'in response to demand from doctors' signalled this change of attitude.)

The RM evaluation found much organizational development going on at RM sites, usually in the form of setting up 'clinical directorates' headed by a hospital consultant. As regards information systems and information technology, many 'case mix' systems were installed under the RM banner. These would contain a database fed from departmental systems, which collected details of patients, diagnoses, treatments and costs. (The Huddersfield clinical information system in the previous case in this chapter could be seen in this light, but its prior definition of expected 'profiles of care', enabling steady learning, made it a superior version.) The actual results in this area revealed the usual disappointments with IT systems: late in implementation, poor outputs, and in some cases 'switched off'. In seven cases out of 150, systems were 'procured but not installed'! (These are the usual results when too early an emphasis on IT neglects prior analysis of the purposeful activity to be supported.)

In terms of the clear concept of RM, as defined in the initiating Health Notice and captured in Figure 4.8, the most significant feature was the amazing rarity of RM projects which *linked* work on information systems (IS) with the relevant organization development (OD). The Health Notice called for clinicians to be 'centrally involved in specifying the information requirements of the new systems'. But on the whole they were not. There were good examples of IS, and examples of good work on OD, but ne'er did the twain meet! This somewhat sad result was confirmed by simultaneous work by a management consultant whose practice is SSM-based, Mike Haynes. He worked on helping RM sites to achieve their required 'sign-off' reports, and found the same gap between the IS and the OD.

Methodologically, a large evaluation involving many people from different professional fields, which could easily have got out of control, was tamed by the guiding principles and devices provided by SSM. These were: the SSM (p) work which led to Figure 4.7; the SSM activity models which shaped both the site visits and the questionnaire to 200+ sites; and the 'RM-ed site' model of Figure 4.8 which enabled coherent analysis of site visit reports. (The entity-based model in Figure 4.8 worked well here because all the NHS sites visited had the same *structural* entities. In general use, SSM's verb-based models are superior, since in human affairs function is more important than form – which is arbitrary and can be easily changed.) The message which all this conveys is clear. Never dive straight into complex situations. Organize prior thinking about intervention by the use of guiding principles about complexity and frameworks and tools for inquiry such as SSM provides.

A more detailed account of this evaluation can be found in Chapter 7 (pp. 198–206) of *Information, Systems and Information Systems*, Peter Checkland and Sue Holwell (John Wiley & Sons, Ltd, 1998).

Case 4: Providing Information Support for Scientists and Technologists

The Situation

In a large, multinational, science-based corporation, with manufacturing plants around the world and central research and development laboratories in Europe, its Information Department (ID) at the laboratories was rethinking the information support it provided to research scientists and technologists. Such provision is always complex, given that in the technology in question the significance of current experiments may become apparent only in five years' time. There is a need both to use known knowledge effectively and to capture new knowledge and make it accessible. The department head decided on an investigation, and asked one of his seniors, Eva, to lead it. She suggested an SSM-based inquiry and asked Peter Checkland and Sue Holwell to help. It was significant that the initiative came from ID, rather than from the researchers themselves. ID was not held in high esteem by the researchers, and the Information Department was anxious to change that.

The laboratories employed 1,000 people; ID consisted of 100. ID contained three technical sections concerned with different aspects of IT and a fourth section (ID1) concerned with managing ID itself and with IS issues. There were, as usual, only a few people who understood clearly the difference between IS and IT. (This is a very normal situation, since the glamour attaches itself to the technology, even though the thinking about 'What information systems shall we set up?' is actually both prime and crucial.) Eva's 'reorganization project', based in ID1, was thus set up to rethink ID's role in providing information support to researchers, using SSM with facilitating help.

The Use of SSM

It was wisely agreed that the reorganization project would include 'clients' of ID from the laboratory's three functional groups: products research; process research; and engineering research. Workshops with about 20 researchers were organized to examine ID/research interactions, with activity models used to structure the work done. Figure 4.9 shows a basic Rich Picture which was developed at the first meeting with members of ID1. It led to definition of relevant roles and processes (listed in Figure 4.9) which the investigation would address. Discussion of this list highlighted one permanent issue of ID's somewhat ambiguous role: Where should the balance lie between the *provision* of information by ID versus *enabling researchers* to help themselves? (This issue is very similar to one raised in Case 3 in Chapter 3.)

As a preliminary structuring device an activity model was made, based on the official mission statement of ID, treating that as if it were a Root Definition. (Now that mission statements are so fashionable, and do at least declare some purposeful activity, it is often useful to model them. Interesting gaps between such models and real-life activity usually appear. This is because mission statements are normally part of PR rather than the result of dispassionate thinking.) The ID mission statement spoke of

> The application of information and IT to laboratory activities, and the integration of IT-based information provision into those activities in ways which are innovative and cost-effective and are a natural part of research planning, both globally and in detail.

The gaps between that concept – expressed in the model shown in Figure 4.10 – and the real situation indicated a strong need for Eva's reorganization project. In the model itself the idea of 'information packages of support' for researchers, developed by ID and researchers working together

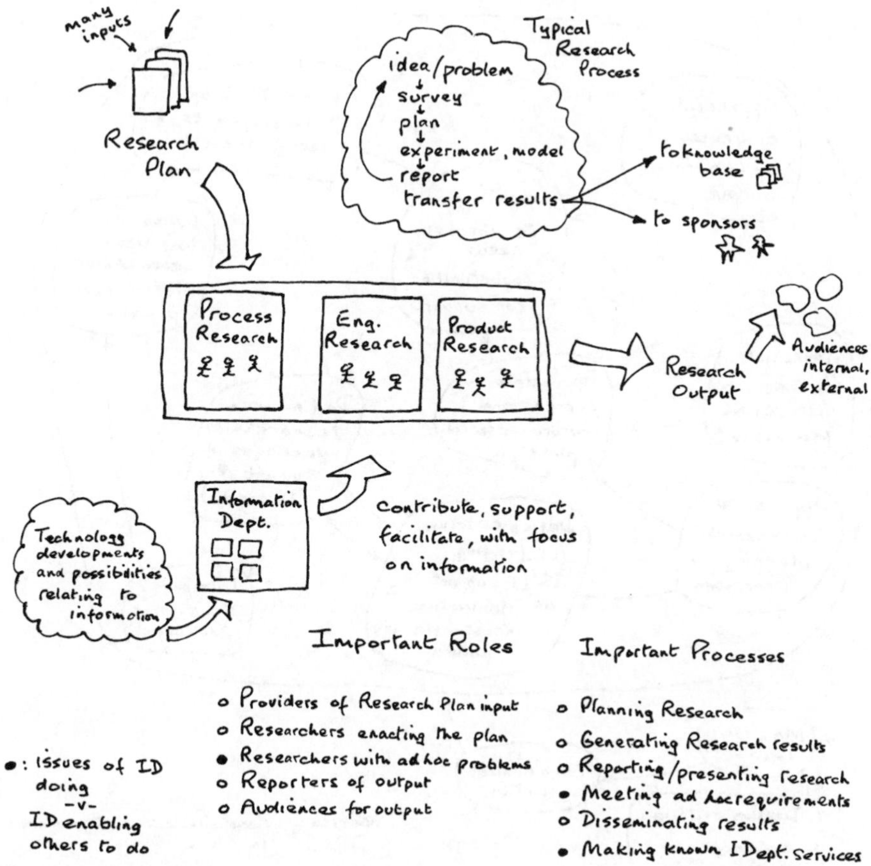

Figure 4.9 An initial picturing of the situation in the research laboratories

(Activity 8) came from the concept of 'integration' in the mission statement. The phrase 'information packages' covered both ID providing information and ID enabling researchers to help themselves. Discussion of this model led to agreement that 'what the researcher needs is more important than how ID wants to use technology' – a small victory for the SSM perspective on information systems.

Two one-day workshops were now designed. These would allow researchers and ID professionals to discuss their interaction and how it might be

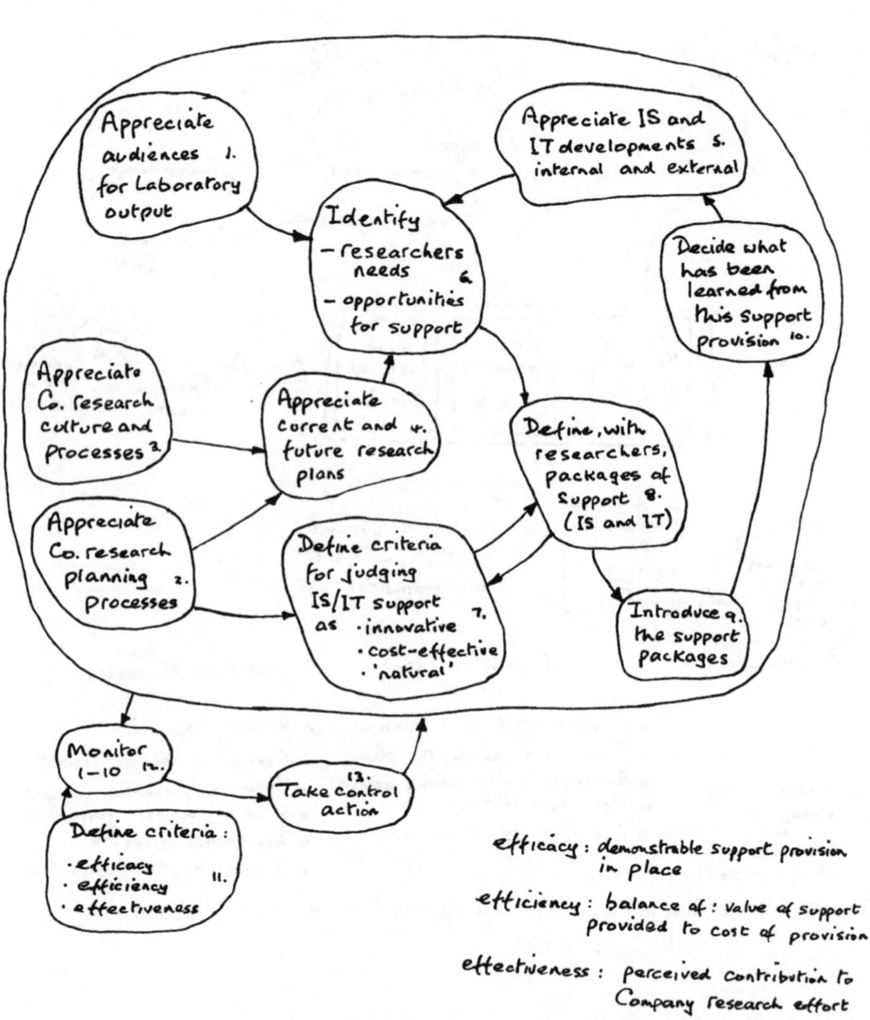

Figure 4.10 An activity model suggested by the Information Department mission statement

developed in the future. At the first workshop, two activity models were used to structure the discussion. These models were relevant to the research laboratories' role in carrying out research and development projects within this company's norms of behaviour. The models were discussed, argued over, modified and commented on in small groups. Their role was to keep the discussion focused while helping researchers and ID gradually to build a shared view of their interaction.

One of the models is shown in Figure 4.11. This is a high-level model relevant to doing research and development in this company in its industry.

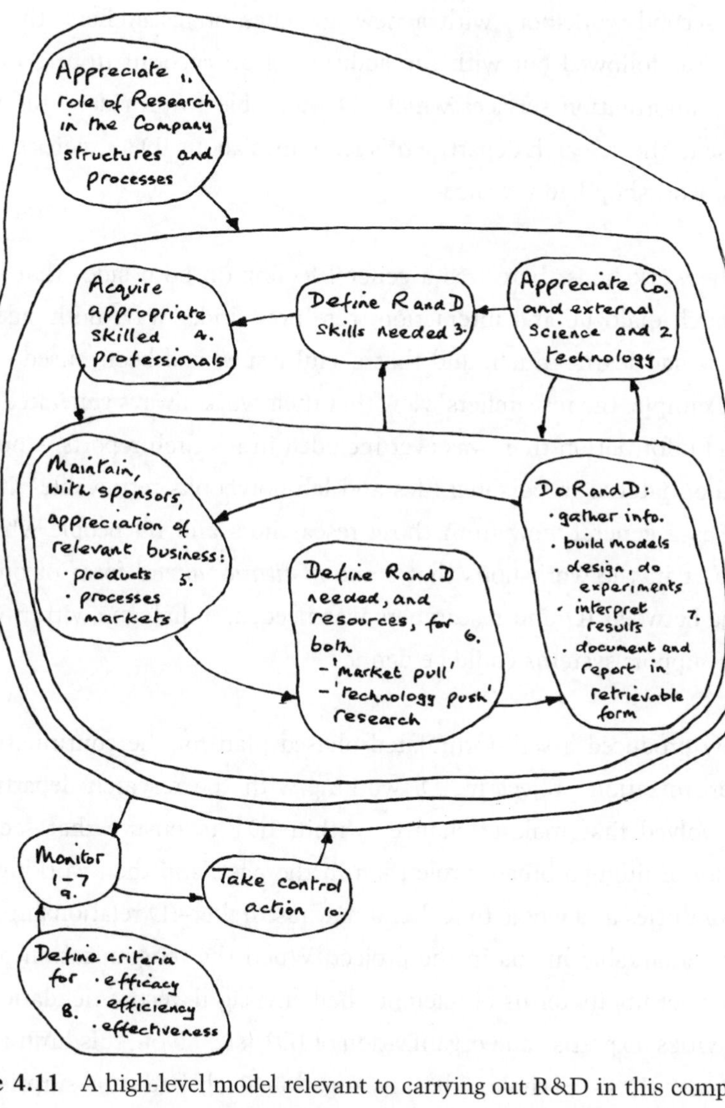

Figure 4.11 A high-level model relevant to carrying out R&D in this company

Another model enabled discussion, activity by activity, of where and how ID could supply helpful 'information packages'. Later discussion in small groups took activities which could in principle be supported by ID and asked of them: Should ID do this; or provide advice, expertise and/or tools relevant to doing it; or provide education and training; or help to manage it?

At the second workshop, with a new audience of researchers, the same pattern was followed but with the addition of an account from ID of the range of information services which ID were able to provide. This was a response to the research department's ignorance about ID's expertise which the first workshop had revealed.

After the workshops there was a general feeling on both sides that useful ID/research dialogue and interaction were now under way. Both sides felt that they had learnt much, and that useful issues had been raised – such as, for example, the researchers' view that their work always generated more data and information than was ever included in research reports. Could ID make the material in personal files and lab notebooks retrievable? Overall there was a general view from those researchers and ID people who had taken part in the workshops that some *institutionalized* form of ongoing dialogue between ID and researchers was needed, a dialogue within which specific support systems could be defined.

Eva now produced a well-formulated phased plan for the continuation of the reorganization project in ID, working with the research department. This involved first making changes within ID1 to ensure that section's capability of filling a broader role than in the past, and then working with the laboratories as a whole to realize a new researcher–ID relationship. Now came a remarkable hiatus in the project. When the reorganization project had been set up, its terms of reference had envisaged 'recommendations on the activities, expertise and organization of ID1 (sic)'. Note this limit to ID1 rather than ID as a whole. The investigation itself had, not surprisingly, taken the thinking beyond ID1 to ID as a whole, and beyond that to the interaction of ID as a whole with the research laboratories. It now became apparent that the Head of ID was daunted by the task of changing not just ID1 but its other, more technically based sections as well. Hence the hiatus as the Head of ID moved forward with great caution – though he did organize an 'away-day' meeting with section heads from the research and development functions to discuss the findings from the workshops.

In fact nearly a year passed, following the workshops, before further progress was made on institutionalizing organized dialogue between ID and its clients. This took an interesting form. An 'Information Market' was set up in the headquarters of the laboratories. This was a permanent physical location at which changing but permanent demonstrations of current IT were displayed. This was described by the head of the laboratories as 'a permanent platform for dialogue'. At the launch of the Information Market, a new Head of general administration – who had returned after a gap spent elsewhere in the corporation – remarked to Checkland and Holwell that he was very pleased indeed to see this development. On his return to the laboratories he had been dismayed to find everything much as he had left it eight years previously.

Outcomes

The outcome of Eva's reorganization project was – eventually – the setting up of organized interaction between ID and its researcher clients via the vehicle of the Information Market. Methodologically the work was straightforward and occupied probably no more than half a dozen days of effort. The organizational change was much much slower. Such disparities are common, and not least in big corporations. Any such organization is bound to be a complex network of roles. The Head of ID occupied one of these, and it was one which gave him a commodity of power by virtue of his role. Also, the company – in Checkland and Holwell's Analyses Two and Three – was rather good at encouraging initiative-taking in role, and also good at encouraging communication between departments, across the network of roles. But the individual concerned – a competent, cautious, if not a charismatic manager – was reluctant to use his power, possibly influenced by the advice he had received from the head of the laboratories when the reorganization project was started: 'Don't focus too soon.' Or perhaps he wanted to find a way of implementing change which would carry the three technical sections of ID wholeheartedly with it – which the Information Market certainly did. That may have been an astute judgement. This remains a particularly interesting example of the interaction of personal issues with broader corporate ones. Never underestimate the complexity of human situations, and do cultivate patience in dealing with them! There is never a straightforward, spick and span answer to the question: What is culturally feasible here?

An account of this investigation is given in Chapter 7 (pp. 174–187) of *Information, Systems and Information Systems* by Peter Checkland and Sue Holwell (John Wiley & Sons, Ltd, 1998).

Conclusion

All the case histories in this and the previous chapter illustrate that using SSM is not like using a formula, cooking to a recipe or painting by numbers. All emphasize that SSM really is a methodology (the logos, the principles of method), not itself a method or technique. Since every human situation is unique, with its own peculiarities, when SSM is used the approach adopted in a particular situation has to be both in tune with the SSM principles and appropriate to the specifics of the situation which make it unique. This use of *methodological principles* to underpin an approach actually used in a particular situation is the process captured in the LUMAS model (Figure 1.8). Each of the cases in Chapters 3 and 4 illustrate that process in action. It is probably significant that the LUMAS model never gets a mention in the secondary literature on SSM. Those who write that literature seem wilfully determined to see SSM as a method rather than a methodology.

Another feature of the cases is that they are all sharply oriented to doing something about a problematical situation. That is always the focus. Once a bit of experience has been gained, the use of SSM is never methodology-oriented. The methodology becomes a given. And it is as experience is gained that the core activity models of SSM (Figure 1.5 and Figure 2.16) become sense-making devices to illuminate the work being done, rather than reminders of a prescription.

5
SSM – Misunderstandings and Craft Skills

Introduction

Having described SSM broadly in Chapter 1, and in detail in Chapter 2, and having illustrated the approach in action in both 'general management' situations (Chapter 3) and in the field of information systems and information technology (Chapter 4), we can now round off the story by covering two further aspects in this chapter. Both should enhance understanding of SSM, one in a rather 'back-handed' way, one more positively.

The back-handed approach reflects the fact that the now voluminous secondary literature on SSM is quite astonishingly full of errors, almost unbelievably so. Some of the most common mistakes will be described here. Our hope is that, in the light of the previous four chapters, it will be completely obvious to the reader that these are indeed gross errors. Knowing this should reinforce understanding of SSM in a positive way. We shall also discuss briefly why these errors occur.

The second aspect covered here is entirely positive. We try to say something useful about what we can call the craft skills in using SSM, skills which develop with experience. We also provide tips from experience in using the techniques within the approach.

Misunderstanding SSM

One of the motivations for writing this book was the publication a year or two ago of a paper by two professors of management science. In a short paragraph purporting to describe the essence of SSM they managed to misrepresent it totally. What was surprising about this was only the source of the errors, not the errors themselves. The fact is, the secondary literature on SSM teems with misunderstandings, so much so that reviewing some of the most common errors serves to reinforce the true nature of SSM as a *process of inquiry* into problematical situations which learns its way to taking action to improve the situations in question. Two examples of the errors in writings about SSM will be given; the first because it comes from an interesting source and is strongly stated, making our disagreement with it very clear, the second because it is so comprehensively wrong.

Before giving the examples, two important points need to be made. The first concerns the tone of the next few pages. In order to demonstrate 'errors' about SSM in the secondary literature we have to do more than simply assert that various statements in that literature are 'wrong'. We have to mount an argument. We have to make clear the claims made about SSM, and then produce evidence to show that those claims are wrong. This calls for a more austere tone in the writing than that which characterizes the rest of this book. We have to operate in academic mode for a page or two. Please bear with us!

The second point ought not to need saying but unfortunately does. Academic discussion is fundamentally concerned with the clash of *ideas* rather than the people who express those ideas. Too often, however, debates among academics becomes personalized and descend to playground levels. We hope to avoid that here.

The first example comes from the book *Complexity and Management* by Stacey, Griffin and Shaw, published in 2000. The book's concern is the theory of organization and management, and it seeks an approach to it that emphasizes 'the self-referential, reflexive nature of humans' and the 'participative nature of human processes of relating'. It aims to find this approach in 'the complexity sciences' when these are 'brought together with psychology and sociology'. In order to realize this project, 'systems thinking' is ditched in favour of 'an alternative to systems thinking about human organizations' which is seen as a 'radical alternative' to the use of systems ideas.

However, when we examine what is being abandoned, it turns out to be the systems thinking of the 1960s. Rejected is 'the very notion that organizations can be thought of either as mechanisms or systems', because thinking like this renders human action subject to 'systemic laws'. Now, anyone familiar with SSM will heartily endorse these ideas! As is obvious from the previous material in this book, breaking away from the assumptions that the world contains a set of interacting systems, and that organizations can be taken to be systems, is the core move in the development of SSM. Its systemicity, its 'systemness', lies in the process of inquiry; the world, and organizations within it, are *not* taken to be systems. It is the (temporary) process of inquiry which can be created as a system, one which learns. However, in *Complexity and Management* SSM is thrust into the 1960s mould. SSM is said to be (present authors' italics) 'a methodology for *systems designers*' and we read: 'Early on in the SSM learning cycle is the requirement for *systemic thinking about the world* and the building of *systems models of that world*.' These statements could not be more wrong. Every time it is referred to, it is wrongly asserted that it is a *system* which is being investigated (and 'designed') rather than a situation. Missing is the appreciation that the models in SSM are only devices to structure a debate about change. Also, the concept of social reality which is accepted for the 'complexity' project turns out to be that which also underpins SSM! Although *Complexity and Management*'s references to SSM literature are not up-to-date (the most recent work referenced being 10 years old), Chapter 8 of the 1981 book on

SSM – Peter Checklands' *Systems Thinking, Systems Practice* (John Wiley & Sons, Ltd) (which *is* referenced) – makes clear how closely SSM's idea of social reality coincides with that in *Complexity and Management*. For SSM:

> social reality is the ever-changing outcome of the social process in which human beings, the product of their genetic inheritance and previous experiences, continually negotiate and re-negotiate with others their perceptions and interpretations of the world. (*Systems Thinking, Systems Practice*, p. 284)

For *Complexity and Management*:

> humans are themselves members of the complex networks that they form . . . With this intersubjective voice people speak as subjects interacting with others in the co-evolution of a jointly constructed reality. (*Complexity and Management*, p. x)

Every one of the case histories described in the books on SSM, including this one, illustrates just this 'co-evolution of a jointly constructed reality' created by 'interacting with others'. That is in fact a very good way of expressing SSM's concern.

The 1981 book on SSM also itself makes it very clear that SSM's models are concepts, rather than descriptions of anything in the world. They are described in *Systems Thinking, Systems Practice* as 'ideal types' in Weber's sense of that phrase; that is to say, 'ideal' in the sense of being 'not physical' (rather than the other sense of 'ideal' meaning 'perfect'). Weber's definition of 'ideal type' is quoted on p. 270, and in the summary of the argument of the book SSM's models are described as:

> 'Intellectual constructions, "ideal types", for use in a debate, *not* attempts to describe reality' (*Systems Thinking, Systems Practice*, p. 19, italics in the original).

Later, in Chapter 8, where SSM's idea of the nature of social reality is discussed, we read that:

> The conceptual models . . . are explicitly 'ideal types' . . . Each model . . . embodies a single one-sided concept . . . a view much purer than the complex perspective we manage to live with in our everyday world. (*Systems Thinking, Systems Practice*, p. 279)

Complexity and Management, an otherwise interesting book, thus offers a compendium of the systems ideas which had to be explicitly rejected in developing SSM: the world as a set of systems, and intervention in the world as a process of designing or redesigning systems.

Two years after the publication of *Complexity and Management*, one of its authors indicated that her thinking had moved on a little from the earlier stance on SSM. In her book *Changing Conversations in Organizations* we read that in SSM: '"systems" are understood to be social constructs, they are not understood as "maps" of any kind of real territory' (pp. 137–138).

Correct! However, according to *Changing Conversations in Organizations*, SSM is concerned with systems not only as SSM models but also as abstractions created in real life by people in a social situation. On p. 138 we read that:

> Soft systems methodologies (sic) . . . view systems as the creative mental constructs of the human beings involved in a process of learning about the divergent ways they are construing their situation.

These 'systems' are taken to emerge as a result of: 'the social practices, culture and politics of those drawing the "system of interest" and developing intentions and making decisions in relation to that system' (p. 138). The phrase 'system of interest' (a phrase foreign to SSM) and the final 'that system' both suggest that SSM sees as systems the social constructions created in dialogue in everyday life. This would mean that SSM's constructed models would be used to gain insight into these 'systems' arising out of social interaction.

This is a very long way indeed from the reality of SSM use, where systems models of purposeful activity are merely the simple devices which enable real-world complexity to be *discussed* in a *structured* way. No assumption that social practice *itself* yields systemic abstractions is made in SSM.

Defining her own intellectual stance, the author declares:

> I am striving to stay with processual thinking which is always incomplete because of the very nature of the dialogically structured conversational realities emerging in reciprocally responsive relationships between living embodied persons. (p. 138)

If we express plainly what we think this remarkable sentence means, we get something like this:

> In human conversation, each of the persons involved influences others and is also influenced by them. Out of this two-way process comes what the participants are creating as their notion of changing 'reality'. These acts of creating reality are never complete, and so have to be examined as only a part of a never-ending process.

Now, this is exactly the view of constructed social reality taken by SSM. SSM's learning cycle is in principle never-ending, since perceptions of the world continue to change. And SSM rejects the language of 'optimizing' or finding 'solutions' to 'problems', for this very reason. As in the discussion of *Complexity and Management* we conclude that SSM, properly used, actually has much to offer the 'complexity sciences'.

The second example is truly astonishing. It was found by Sue Holwell in her doctoral research on the development of SSM. It is from a 1995 textbook on information systems, in which we read, concerning SSM:

> This methodology stems from the work of Checkland (1981) who took a radically different approach *to the analysis and design of information human activity systems.* Starting from the premise that *organizations (and therefore their subsystem information systems) are open systems that interact with their environment,* he *includes the human activity subsystems* as part of his modelling process. The methodology starts by taking a particular view *of the system* that includes the people involved, the problem areas, sources of conflict and other 'soft' aspects *of* the overall system. A 'root definition' is then formed *about the system* which *proposes improvements* to *the system* to *tackle the problems identified in the rich picture.*
>
> Using *the* root definition, *various* conceptual models *of the new system* can be built, compared and evaluated against the *problems in the rich picture. A set of recommendations* is then suggested to deal with the specific changes that are necessary to *solve the problems.* These are evaluated in terms of feasibility and used to propose specific remedies for action. (Present authors' italics, which indicate errors)

This manages to cram about 20 errors into fewer than 200 words. SSM is *not* concerned with 'the analysis and design of information human activity systems'. It does *not* take organizations to be systems which have information and human activity subsystems. It does *not*, at the start, take a particular view of 'the system'... There is no need to go on. This example, as this book shows, is almost magnificent in its wrong-headedness.

We have given here two illustrative misunderstandings of SSM out of scores which could have been chosen, and it is interesting to speculate as to why it is that failing to understand SSM is such a common phenomenon. Its ideas are not abstruse; understanding it is not like trying to understand quantum mechanics or cosmology. A general explanation may be that many people, reading casually, pick up a few trigger words or phrases and then read into the text what they expect to find there. In this case the word 'system' could act as just such a trigger.

The situation appears to be this. As was stated in Chapter 1, *in everyday language* a number of more or less complex chunks of reality are referred to as 'systems'. We speak of 'health-care systems', 'the transport system', 'the economic system', etc. None of these chunks of reality actually meet the requirements of the word 'system' when that is carefully defined in the course of serious scholarly work – such as in Figure 1.1 in this book. However, the everyday-language use of the word is now so embedded in our heads and culture that many people, as soon as they read the word 'system', automatically assume that it refers to something in the world, and this cuts off any other strand of thought. Unconsciously, they are then adopting the outlook underlying the 'hard' systems thinking of the 1960s, and they cannot thereafter either hear or consider the outlook which abandons that position, and instead sees systemness as being applied to the process of engaging with the world (as in Figure 1.7) rather than in the world itself. Thus it becomes difficult for many people to hear, let alone embrace, the argument for treating the real world as an exceptionally complex mess, but one which can nevertheless be coped with by a process of inquiry organized as a 'learning system'.

In SSM the word 'system' never refers to any entities 'out there' in the world. However, it does have two legitimate places within the methodology and its use. First, the concepts used as devices to structure discussion about change happen to be constructed as *purposeful activity systems*. (This is because the messy real world does contain a lot of interactions and connections and much would-be purposeful action.) And second, in methodology use, the practitioner should strive to ensure that the process of inquiry which leads to action is created as a *learning system*. (This is because there are no final answers in human affairs, learning being ongoing and permanent.)

Craft Skills in SSM Use

It is not easy to talk or write clearly and explicitly about craft skills, for the phrase conveys the idea of something which cannot be pinned down explicitly, something rather mysterious which cannot be completely analysed. Craft skills can be acquired through experience but cannot entirely be taught, not in the way that 'How to solve simultaneous equations' or 'How to set up a website' can. The response to this in everyday life is the idea of 'apprenticeship'. The would-be young potter works alongside the skilled potter and eventually may be able to produce high-quality pots, having absorbed much from his or her mentor in terms of both explicit and tacit (unexpressed) knowledge. Now, we are not claiming that apprenticeship is necessary to become a competent SSM practitioner. There are many examples of people who have made excellent use of SSM based on written accounts of it. But we use this example to illustrate the fact that the process of using a methodology is much richer than the biff-bang application of a technique. What we *are* claiming is that with experience the user of SSM will both find a way of using the methodology that they are personally comfortable with (which fits with their cast of mind) *and* improve their use of SSM as experience accumulates.

Meanwhile, we can offer some advice from experience which may help with the process of internalizing SSM, so that attention can be directed wholly to the situation addressed, rather than addressed to the methodology. Progress in that is signalled by no longer having to ask such questions as: Remind me again, what was the difference between Primary-Task and Issue-based Root Definitions? Get over that hurdle and you can really begin to use the methodology effectively. In fact worrying about the methodology or its tools can hinder the learning process. The best advice about SSM is: dive in, tackle real situations and learn about SSM along the way.

The craft skills in SSM use are thinking skills, rather than physical skills, and so can be thought about while sitting at a desk, going for a walk or lying in the bath. Here, from experience, are some remarks about the practitioner state of mind which will make it easier to develop SSM's craft skills.

1. Always remain conscious of the fact that the process in which the user of SSM is engaged is the one of addressing a complex human situation, mentally, by the conscious organized use of particular ideas and principles in order to achieve sense-making, as shown in Figure 5.1.

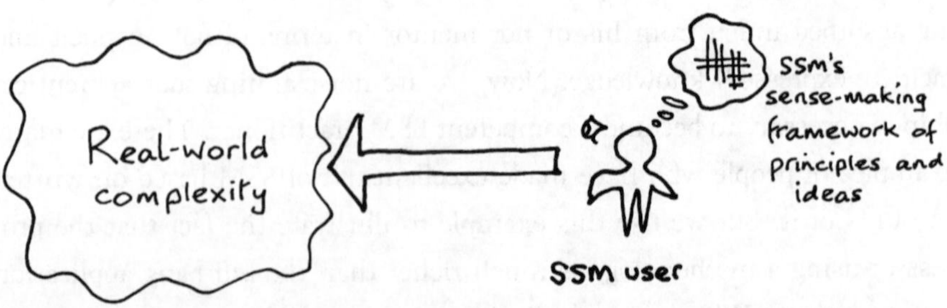

Figure 5.1 SSM's basic stance – using a particular set of principles and ideas to make sense of real-world complexity

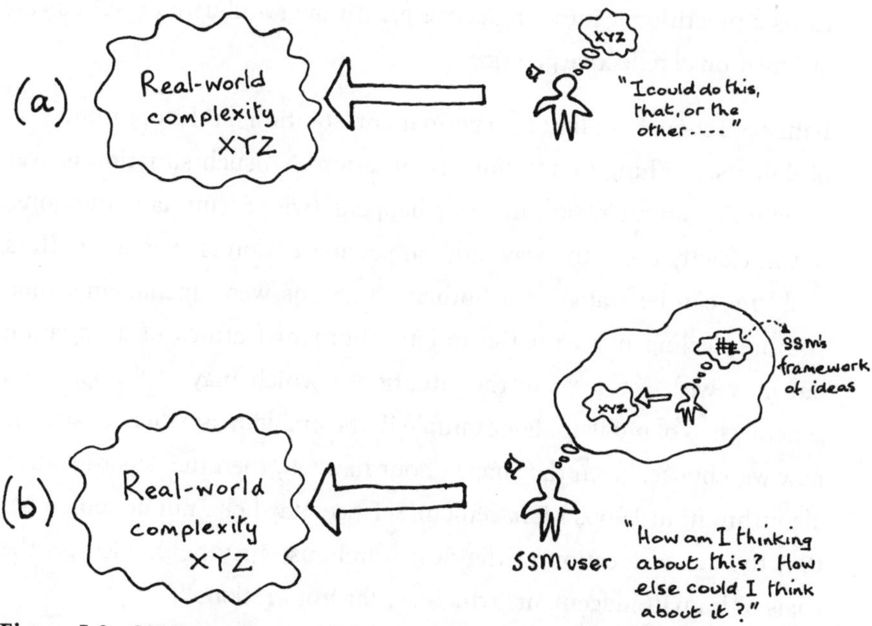

Figure 5.2 SSM, as reflective practice, entails consciously thinking about your own thinking, i.e. moving to stance (b)

This implies what is probably the key step in really understanding SSM and its use: grasping that the user in Figure 5.1 is *consciously thinking about his or her own thinking*. This 'meta-level' thinking is not all that common. Some extremely intelligent people go through life in the stance shown as (a) in Figure 5.2, never thinking about themselves as thinkers. They perceive complexity in the world outside themselves and, *at the same mental level as that perception*, think 'I could do this, that, or the other . . . ' The experienced SSM user is in the stance shown as (b) in Figure 5.2. This lifts the thinking to a level above that of simply perceiving the complexity. It lifts it to a meta-level, and makes the user able to inspect their own thinking and then think about it. It is this shift from stance (a) to stance (b) which increases the richness of thinking and enables insights to emerge and formula-driven thinking to be avoided. It is the (a) to (b) shift which turns a practitioner into a reflective practitioner and defines SSM as an articulation of reflective practice.

2. Banish all thought of finding a permanent 'solution' or the optimum way of doing something in any human situation. No such situation is ever *exactly* like another; nothing ever happens twice in human situations, not in exactly the same way, and no such situation is ever static. (If 'a problem' can be stated *as if* human situations were unchanging, then you are dealing not with the unique [human] features of a situation but only with the logic of the situation – which may well apply to a general class of problem. For example if 'the problem' is 'where to site the new warehouse, given the shape of our market', then the depot-location algorithm from 1960s Management Science may help. But do remember that the actual location of the new warehouse may be decided on the basis of human judgements which are far from rational.)

3. Try not to *impose* a structure on the situation. Rather let it 'speak to you', as you tease out the strands of thinking within it. The attitude to adopt is that implied by the Scottish phrase, 'I hear you.' This means withholding judgement, neither approving nor disapproving of what you find but allowing the situation to reveal its patterns. And know that this pattern can (probably will) change within the course of an investigation. So, be positive in forming judgements about the situation but tentative about hanging on to those judgements. Also, revisit your thinking continually to see how both the situation and your thinking about it is changing.

4. Remember that no methodology can do your thinking for you, and lead inevitably to a unique and successful outcome. What it can do is structure your thinking, or that of a team carrying out an investigation, so that your and/or the team's capabilities are used to the full. Also, in the team case, it will *enable* a group of people to become a real team much more easily than would be the case if no declared methodological principles were being followed. In virtually all the case histories in Chapters 3 and 4,

SSM acted in this way, providing shared concepts and a shared language which helped team coherence. In another example, not described here, a team of civil servants and outside consultants carried out an SSM-based study of the personal taxation arrangements in the UK. An SSM (p) model of 'a system to do the study' was built at the start, based on the study's terms of reference, which were treated as a Root Definition. It was used continually as a sense-making device as the study unfolded (rather than as a plan), and ensured that there were no communication issues in the disparate but united team.

5. When facilitating an investigation being carried out by people in the situation, always keep in mind the principle that your aim is to *give away* the approach being used to the people in the situation itself. Don't hang on to ownership. In the rethinking of the role of Shell's manufacturing function (Chapter 3) the investigation was truly carried out, with facilitating help, by the participants in the workshops, not only by the facilitators. In the rethink of the Information and Library Services Department (Chapter 3) the three members of the department who were seconded to carry out the study part-time wished to give the internal presentation on the finished work without help from the facilitators. This was an important signal that the higher-level aim of the study (to increase the department's 'problem-solving skills') was being achieved.

6. Be ready to be surprised by the turns which the investigations take. As worldviews are surfaced and questioned there is no knowing which way the work will go, or what the final outcome will be. In the work in the publishing and printing corporation (Chapter 3) the outcome of structural change was in no way envisaged at the start of the work. Outcomes derive from no formula, they arise from the idiosyncrasies of the situations addressed. They derive in part from the always-present tension between the glorious mix of altruistic behaviour directed to group aims and the selfish pursuit of personal agendas which is never absent from human affairs.

7. Be aware that the action emerging as desirable and feasible from an investigation will frequently not be implementable by those undertaking that study, who may not have the necessary power. Because of this the investigators need continually to be making judgements about possible outcomes and asking themselves who would be in a position to cause or authorize action to be taken. Then make sure that those people are as closely involved as possible in the course of the investigation. It may not be possible to draw them into participation (they may well be senior people with wide agendas and full diaries); but as a minimum make sure that the outcome of an investigation does not come as a big surprise, out of a clear sky. Take whatever enabling action is necessary to avoid that.
8. Don't let the work done as part of an investigation ever feel like 'work', grinding along under grey skies. If it does feel like a grind, rather than an intellectual adventure, then stir things up. Try some outrageous Root Definitions, redefine CATWOE elements, think of new possible (if improbable) 'issue owners' in Analysis One. Do whatever you have to do to recapture zest. SSM use should never feel like a grim or plodding experience; it should always be fun, serious fun, and a rewarding experience.

Tips on Techniques in SSM

This section completes the chapter by giving some advice on the use of the various techniques within SSM, together with some general tips on facilitating an SSM-based investigation.

Finding Out

1. Get into the habit of continually making rich pictures, or fragments of rich pictures, remembering that the aim is always to capture the key *relationships* in the situation of concern. Use pen and paper rather than

a computer screen since fiddling with the technology seems to get in the way of exercising the imagination.

2. Always use the list of issue owners from Analysis One (Figure 2.3) as a prime source of ideas for relevant activity models.

3. Accept that in most organizations *open* discussion of Analysis Three's insights on power will not be achievable. Pursue this in one-to-one or small-group discussion. And always supplement Analysis Three by finding out what the current in-house organizational jokes are. They usually relate to issues of power.

Model Building

4. Any purposeful activity which is to be expressed as a model should be thought about initially as an input–output transformation, T. Then expand it into a model by use of PQR, a full Root Definition, CATWOE and E_1, E_2, E_3.

5. Since models should be built very literally from the words used in the Root Definition, etc. avoid metaphors in RDs. (Checkland, externally examining Masters dissertations at a technical university, once came across a Root Definition related to a small firm which provided services to its clients. It began: 'A system to go the extra mile for our clients'! Not a good idea in these words if you wish to model this activity.)

6. Make sure that the activities you include in a model are feet-on-the ground activities which someone could in principle go away and do. Avoid naming activities such as 'create a learning organization'. If you felt you must use that fashionable but highly ambiguous term, put in the model the activities you think you would have to do in order to achieve an organization which might merit that description. Also, keep as far as possible to the '7 ± 2' guideline for the activities in the operational part of a model. This ensures that the model's activities are broadly at the same

level. Individual activities can then become a source of more detailed models if necessary.

7. Always try to work with both issue-based and primary-task root definitions. And make sure that there is always a plausible link between the O of CATWOE and E_3, since the notional 'owners' of this purposeful activity, if it were to exist, would be interested in the higher-level or longer-term aim of the activity system, as measured by the effectiveness measure, E_3.

8. The most recent pattern to emerge in large complex investigations is the practice of making an SSM (p) model of 'a system to carry out the investigation' at a point at which the major themes have emerged. Such a model is then used as a sense-making device (rather than as a plan or prescription). Used continually, though not continuously, such a model enables the investigators to step back and appraise how the investigation is going.

Using Models to Structure Discussion of Change

9. Experienced users of SSM, before building a model, will always examine a number of possible models *mentally*, asking themselves: If this idea for a Root Definition were to be turned into a model, what kind of 'comparison' with the real world would it lead to? Remember, also, that a model's CATWOE can be used in its own right as a source of questions to ask of the real situation. This can help to pinpoint models worth building, namely those which stimulate interest and catch attention.

10. Always think in terms of finding accommodations that will enable '*change* which will improve the situation' to be defined, rather than in terms of 'a system to be implemented'. The experience of several hundred investigations shows that implementation of a system is the rare special case of 'change', usually at an uncontroversial tactical level – for example, in a big mail order company, implementing a system to

deal effectively with the incoming mail each morning. Nobody will argue against that.

11. Never get bogged down in filling in the whole of a 'comparison chart' in agonizing detail. First glide gracefully over the chart in discussion, in order to get a feel for where the issues and energy lie. Then be selective.

12. When ideas for feasible and desirable change begin to emerge, remember that it is always possible to sharpen them by thinking of 'a system to *implement this change* in this organization and situation'.

13. Always allow as much time as possible for the sense-making process to work. Real-world complexity can never be grasped instantly. Similarly any decision to be taken will benefit from as long a period of reflection as can be arranged.

Facilitation

14. Although it is possible for an individual to carry out an investigation using SSM, the ideas are most fruitful when used in group activity. Record what happens – for example on flip charts – and later document the course of the study in more reflective documents. Date each flip chart and each document and copy this documentation to all members for their reflection and discussion. In this way the history of the changing thinking can be recovered subsequently. This will enhance learning from the experience. Beyond that, all the usual features of facilitation apply to an SSM investigation: encourage participation from all of those present, and create spaces in any schedule so that there is time for the thinking to soak in. In Guy Claxton's phrase, this helps to enable the 'tortoise mind' – in which serious thinking and sedimentation of ideas occurs – to catch up with the 'hare brain' by which we normally live our lives from minute to minute.

Part three
Summing up

6
SSM as a Whole – Some Reminders

The account of SSM and its use is completed in the five previous chapters. This short finale provides some reminders of the whole, as an aide-memoire. First we suggest some features of the appropriate mindset when approaching SSM, then provide a diagram summarizing it as a whole.

Approaching SSM: The Mindset

It is not very usual in Western thought to devote much attention to thinking about thinking. In most subject areas the focus is always on the substantive content, while 'how to think about this' is neglected. It is assumed that serious attention to the subject matter will somehow also inculcate 'how to think about it' by some kind of osmosis. However, the output generated by SSM's 30-year programme of action research in problematical real-life situations was precisely *an explicit way of thinking about*, and hence a process for dealing with, the kind of complexity found in human affairs. (The unusual nature of this outcome is probably what can make it difficult for some people to understand it.) The nature of SSM as a methodology implies a particular view of social reality (which is summarized in Appendix A) but also implies that a would-be user should approach it in a particular frame of mind, which is summarized here in seven pieces of advice.

- *Reflect* on the fact that most discussion in human situations is of poor quality. Different topics interact, participants speak at different levels (from tactical to strategic) and bring different judgements to bear, based on different (unacknowledged) worldviews.

- *Know* that SSM can make such discussion much more coherent, and will deepen the level of thinking due to its surfacing of worldviews, since these govern how issues are both perceived and judged.

- *Accept* that no methodology can on its own lead to some first-rate outcome, but know also that even rough-and-ready use of SSM will improve the quality of the thinking of the participants and increase the quality of the discussion which they generate.

- *Know* that methodology should be treated for what it is, a set of principles which need to be tailored to a method for *this* situation with these participants, with their history, now. And remain oriented to the problem situation, not to the methodology, using SSM for making sense of real-life complexity.

- *Know* that the best way to learn about SSM is to use it, however crudely you do this at first.

- *Know*, when having a go at using it, that its principles are very resilient, capable of standing up to a good deal of rough use. (Models which might not get high marks in a university exam can, in real life, be helpful!)

- *Know* that the understanding of a situation gained through use of SSM is not gained for its own sake, but to become a spring for action. This is an action-oriented approach.

Given the frame of mind outlined above, any problematical situation in human affairs may be tackled with some confidence.

The outcome of any use of SSM will depend upon a number of factors whose effects cannot easily be disentangled. These include: the characteristics and abilities of the people carrying out the investigation; the characteristics of the situation as perceived by those who care about it; and the methodology itself. To this we can add that the very best uses of the approach seem to exhibit a certain style. At the end of the 1990 book describing 12 uses of mature SSM (Peter Checkland and Jim Scholes, *SSM in Action* [John Wiley & Sons, Ltd]), this was stated in the following terms:

> the very best uses of SSM seem always to exhibit a certain dash, a light-footedness, a deft charm. In this sense the role of the approach is akin to that of the cavalry in nineteenth century war: it can add a certain tone to what might otherwise be a vulgar brawl. (p. 302)

To this we can add that the confidence which comes from SSM once it is internalized can help you, in the midst of the turmoil of everyday life, to demonstrate one highly desirable and productive end: grace under pressure.

The final page of the main part of this book is a one-page aide-memoire of the key elements in SSM's learning cycle, and their relationships within the whole. Good luck with it!

A Basic outline of Soft Systems Methodology

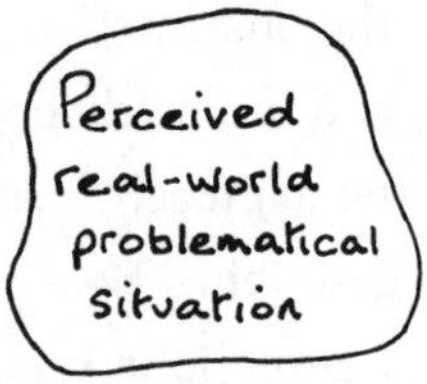

- Think: 'problem situation' not 'problem'
- Find out about it:
  Rich Pictures
  Analysis One (the intervention)
  Two (social)
  Three (political)

- Think of some 'relevant systems' of purposeful activity; name the worldviews they encapsulate

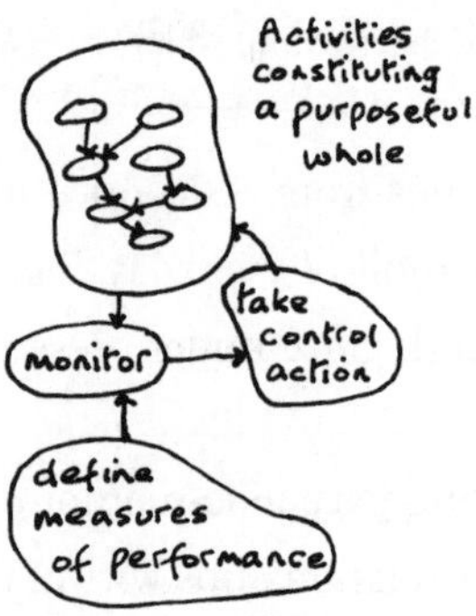

- Build the models of these notional systems (as on the left)
  - Root Definitions (PT/1B)
  - PQR
  - CATWOE
  - 3Es efficacy
    efficiency
    effectiveness

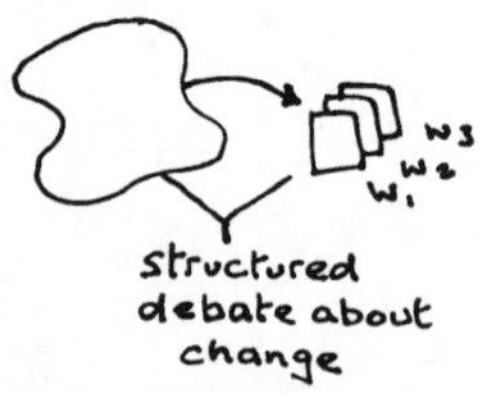

- Use the models to question the perceived real-world situation, structuring a debate about change

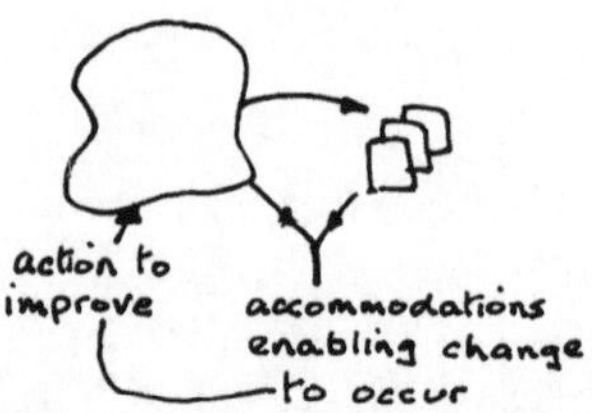

- Seek accommodations which meet criteria: systemically desirable (based on these models) and culturally feasible (for these people in their situation). An accommodation is a version of the situation which different people (different worldviews) can nevertheless live with.

Appendix A
SSM's Theory

It is not impossible to be a good SSM practitioner without knowing about, or being interested in the theory which underpins it. However, it is more than likely that understanding the theory will have a positive effect in improving any practitioner's use of the methodology. Appendix A therefore gives an outline of the (social) theory which is consistent with SSM, and indicates where more detailed accounts can be found. It covers both the theory underlying the SSM approach itself and the theory underlying how it is used.

The Theory behind SSM

Whenever anyone acting in good faith takes some deliberate, intentional action which they regard as 'sensible', behind the action there must be some ideas which lead to that particular action taking place. These are the ideas which define what counts as 'sensible' for the person in question in the situation in which they find themselves. These abstract ideas constitute a theory which justifies the action. Thus we are all in one sense theoreticians whenever we take intentional action, even though the ideas in question may not have been carefully thought out, and may not be expressed. The great economist Keynes remarked that when he talked to businessmen they usually claimed that they were down-to-earth practical people, with no time for airy-fairy economic theory. But, he said, they usually turned out to be prisoners of some out-of-date theory from 30 years ago! This tale both illustrates that practice is always linked to theory, and that all theory is in the end provisional, and may be replaced as new experience accumulates.

In similar vein, any process for intervening sensibly in real-world situations to bring about 'improvement' must have some ideas – some theory – about the nature of social reality, whether it is made explicit or not. There must be some theory which makes any chosen process of intervention 'sensible'.

During the Second World War much useful experience was gained from the practice of including scientists in operational groups. In the post-war period this led to a rapid development of so-called 'management science'. By the end of the 1960s a number of approaches to tackling real-world problematical situations had matured: systems engineering, classic Operational Research, RAND Corporation systems analysis, computer systems analysis, System Dynamics, the Viable Systems Model, etc. Interestingly, all these approaches (now thought of as 'hard') assume essentially the same theory of social reality, although their literature does not usually discuss this aspect of them. What has happened since these methods became mature is that further experience has shown that their social theory is not rich enough to encompass the astonishingly florid complexity of human situations. New approaches (now thought of as 'soft'), underpinned by a different social theory, have emerged. They do not, however, suggest that the 1960s theory was 'wrong' and should be abandoned. Rather the 'new' theory sees the 'old' one as a special case, perfectly adequate in certain circumstances but less general than the social theory behind the 'soft' outlook.

So what was the image of social reality behind the 1960s approaches, and what more general social theory characterizes an approach such as SSM?

At a practical level the 1960s approaches all take completely seriously the ability of human beings to define precise objectives and then to organize activity to achieve the optimum state of: 'objectives achieved'. The image is of human beings (and organizations) as *goal seeking* and *optimizing*. These ideas do of course adequately describe some important aspects of human activity. But the failure to transfer the Systems Engineering approach from technically defined problematical situations to management situations led, in the development of SSM, to the realization that richer and broader language was required to make sense of the SSM experiences. Eventually, the notion of seeking objectives was subsumed in the broader concept of *sustaining relationships*. This contains goal seeking as a special case. (Thus if a company has 'the objective' of increasing its market share by means of its new product, this is a special case of that company creating and sustaining a new relationship between itself and the market.) Similarly, the notion of engineering an optimum outcome was subsumed in the broader concept of *learning*, which itself contains optimizing as a special case.

This shift from goal seeking/optimizing to sustaining relationships/learning turned out to be a much bigger shift than was at first realized. It is, in the language of academic social theory, a shift from one philosophy and sociology to a different philosophy and a different sociology. It is a move from positivism and functionalism (the 'hard' approach) to phenomenology and interpretive sociology (the 'soft' approach). The shift is discussed in these academic terms in Chapter 8 of the first book on SSM, *Systems Thinking, Systems Practice* (by Peter Checkland, John Wiley & Sons, Ltd, 1981). That will not be pursued here, except to indicate that the nature of the shift is to move away from a static view of social reality (ignoring worldviews) as something 'out there' which can be studied objectively by an outside observer as if social reality were similar to natural phenomena, to a process view (encompassing worldviews) which sees social reality as something continuously being constructed and reconstructed by human beings in talk and action. That is the shift from 'positivism' to 'phenomenology'. The adoption of this latter view was essential if we were to make sense of the experiences in human situations which comprised the development of SSM. It implies a dynamic view of social reality like that shown in Figure A.A.1. This model provides a language in terms of which SSM makes sense; it is a version of the social theory which lies behind SSM.

Finally, it is interesting to note that recent developments in both Operational Research (OR) and System Dynamics (SD) suggest that the same shift of theory which had to be made during SSM's development is also taking place in those fields. 'Soft OR', comprising such approaches as SSM, SODA (Strategic Options Development and Analysis) and SCA (the Strategic Choice Approach), is now accepted as a significant development of OR. And in the System Dynamics field the notion of 'qualitative SD' – using SD techniques to structure inquiry – is increasingly prominent. Both developments represent a move from positivist to phenomenological social theory as their underpinning.

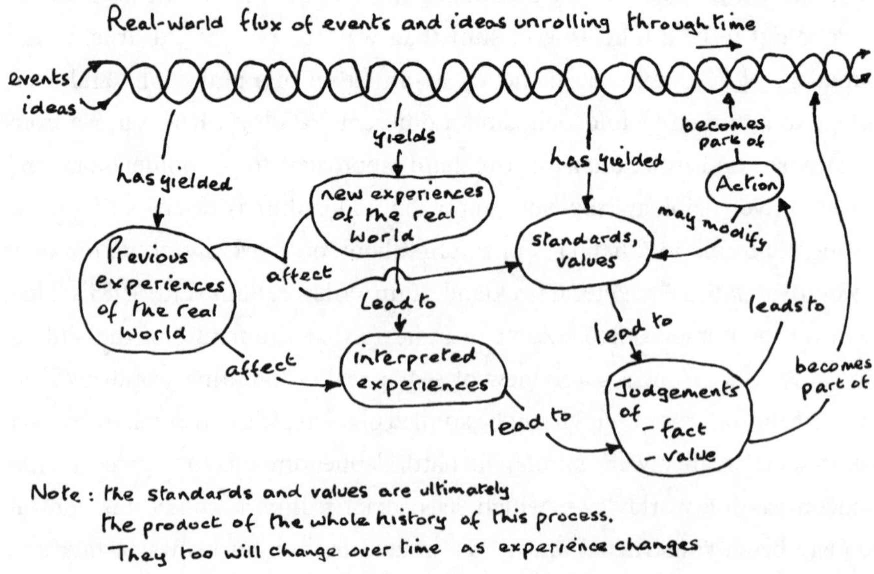

Figure A.A.1 The dynamic social process: a model which makes sense of SSM use

Further Reading

- For discussion of the theory of social reality embodied in SSM, see Chapter 8 of *Systems Thinking, Systems Practice*, by Peter Checkland (John Wiley & Sons, Ltd, 1981 and 1999) (the latter containing also *SSM: a 30-year retrospective*).
- For discussion of the hard/soft distinction, see Peter Checkland and Sue Holwell, 'Classic' OR and 'Soft' OR – an asymmetric complementarity, Chapter 3 in M. Pidd (ed.) *Systems Modelling: Theory and Practice* (John Wiley & Sons, Ltd, 2004).
- For the background to Figure A.A.1 in the work of Vickers, see: Systems theory and management thinking, Appendix to *Soft Systems Methodology: A 30-year Retrospective*, by Peter Checkland (John Wiley & Sons, Ltd, 1999), included in paperback versions of both Peter Checkland, *Systems Thinking, Systems Practice*; and Peter Checkland and Jim Scholes, *SSM in Action* (John Wiley & Sons, Ltd 1999).

The Theory behind SSM Use

Since every use of SSM tries to be an organized process of *learning*, there is a sense in which every use of the approach in practice is also a piece of *research*. This was of course explicitly the case as SSM was being created out of real-world experiences.

But even with SSM now mature, the process in the LUMAS model (Figure 1.8) still applies, and no two uses of SSM ever follow exactly the same path. Every use entails learning, so every use is, in that sense, research.

If we look at the use of SSM in this light we find that the theory of research underpinning it is not at all the conventional theory which dominates academic discussion and practice, namely research conducted as hypothesis testing. This domination is understandable, since hypothesis testing is the research method in the natural sciences, and these embody the most successful way of finding things out that the world has ever seen. In natural science you design and set up an experiment to test a hypothesis. If the results are *repeatable* by other scientists in other locations at other times, then the results count as part of scientific knowledge. Repeatability is a very strong criterion for truth. And progress – of a curious backward kind – can be made, based on this criterion, in a sequence of experiments in which attempts are made to refute hypotheses. Scientific knowledge then consists of the mass of hypotheses which have passed severe tests and have not (so far) been refuted. This works because natural phenomena are not capricious; they are (above the quantum level) unchanging while being experimented on. If you drop a piece of sodium onto water, it fizzes as hydrogen is released and the water becomes a solution of sodium hydroxide. It does not matter who does this, where, or when; the result is always the same. This is a part of testable scientific knowledge.

When we turn to *social* phenomena, however, the picture is much more complex, and how to research them is much less clear. Social phenomena *are* capricious. They are necessarily expressed in abstract terms, and are subject to multiple and changing interpretations. The phenomena change as interpretations change over time. (Consider the slipperiness of concepts like 'educating' someone or ' managing' something.) No body of social knowledge comparable to the accumulated knowledge of the natural sciences has been produced.

A radical approach to researching in social situations was suggested by Kurt Lewin, a German psychologist who emigrated to the United States in the 1930s. Given the impossibility, in social research, of remaining an outside observer of an unchanging phenomenon, he suggested researching by entering a social situation, *taking part* in the action going on (in whatever direction it goes) and using the involvement as a research experience focused on the change process. This so-called 'action research' was the research method adopted in the development of SSM.

The great issue with 'action research' is obvious: What is its truth criterion? It cannot be the 'repeatability' of natural science, for no human (social) situation ever exactly duplicates another such situation. So if an action researcher gives an account of an action research project in a report, how is this different from story telling or novel writing?

The solution to this dilemma is to use a criterion which is necessarily less strong than 'repeatability' but is stronger than the 'plausibility' which is all some sociologists aim for. The criterion is 'recoverability'; that is to say, make the whole activity of the researcher (here the SSM user) absolutely explicit (including the thinking as well as the activity), so that an outside observer can follow the whole process and understand exactly how the outcomes came about. If the observer then wishes to disagree with the actions taken or the interpretations made, coherent discussion can take place. For this recoverability to be achievable, however, a particular condition must be fulfilled: the researcher (or in our case the SSM user) must state in advance of undertaking the research/investigation, the framework of language (the epistemology) in terms of which what counts as knowledge from the work will be expressed. For the SSM user the methodology provides exactly such a framework of language, with many carefully defined terms — Analysis One, Two and Three, Root Definition, CATWOE, PQR, etc.

Thus, the theory underpinning SSM use is that of action research, with particular emphasis on the advance declaration of the language in terms of which knowledge will be defined. This is done by declaring the use of SSM at the start of an investigation. It is very unfortunate that the now extensive and rapidly growing literature on action research is rather poverty-stricken, with little attention being addressed to making sure its findings are in some sense valid, even though complete repeatability can never be achieved.

Further Reading

- For 'hard' and 'soft' views of 'an organization', and discussion of action research, see Chapters 3 and 1, respectively of *Information, Systems and Information Systems*, by Peter Checkland and Sue Holwell (John Wiley & Sons, Ltd, 1998).
- For discussion of action research and the 'recoverability' criterion, see Peter Checkland and Sue Holwell, Action research: its nature and validity, *Systemic Practice and Action Research,* Vol. 11(1), 9–21, 1998.

Appendix B The Story of SSM's Development

The first paper on what became SSM appeared in 1972 in the *Journal of Systems Engineering*: Peter Checkland's 'Towards a systems-based methodology for real-world problem solving' (Vol. 3, No. 2). The basic shape of SSM had by then emerged, but the flavour of the paper is much closer to its origins in systems engineering than it is to mature SSM. The crucial shift from 'hard' to 'soft' systems thinking had not then been fully recognized. Since then, the story of SSM's development and use has been told in many more papers and in the following four books.

- *Systems Thinking, Systems Practice* Peter Checkland (John Wiley & Sons, Ltd, 1981, 1999) This book recounts briefly the emergence of the method of scientific inquiry, which is based on reductionism, repeatability and the refutation of hypotheses. It then covers the emergence, out of biology and engineering, of the systems movement as a would-be counter to the reductionism of natural science. It then shows how SSM emerged through the idea of treating purposeful activity as a systems concept. Six early investigations are described, and the final chapter discusses the implications of these, establishing the view of the nature of social reality which SSM entails, namely that social reality is continuously socially constructed and reconstructed.

- *Soft Systems Methodology in Action* Peter Checkland and Jim Scholes (John Wiley & Sons, Ltd, 1990, 1999) The first chapter places SSM in the context of 'managing' as a process, and establishes the 'four actions' model of SSM which has become the norm (Figure 1.5 in this book). Chapter 2 describes the then 'developed form of SSM' and later chapters describe 11 investigations, covering situations in industry, and in the UK's National Health Service and Civil Service.
- *Information, Systems and Information Systems* Peter Checkland and Sue Holwell (John Wiley & Sons, Ltd, 1998) The focus here is the conceptually confused field of information systems – the book's subtitle being: 'making sense of the field'. Part One surveys the field's confusion; Part Two uses systems ideas to establish a clear concept of the field's core, namely the provision of information to support purposeful action. Part Three describes the development through action research of one of the most important information systems ever created – the system (pre-dating computers) which converted radar data into fighter-control information, enabling the Royal Air Force to win the Battle of Britain in 1940. The action research approach is described, followed by accounts of SSM-based investigations in both industry and the National Health Service. Part Four describes models teased out of these experiences, which provide a unified concept of the field.
- *Soft Systems Methodology: A 30-year Retrospective* Peter Checkland (John Wiley & Sons, Ltd, 1999; included in the paperback editions of both the 1981 and the 1990 books. This retrospective reviews the whole history of SSM development, introduces the LUMAS model (Figure 1.8 in this book) and in an appendix places SSM as a way of operationalizing Geoffrey Vickers' view of the social process as 'an appreciative system' (from which Figure A.A.1 in this book derives).

Appendix C
A Talked-through Example of Activity Model Building

As an illustration of the process of building a model of a purposeful activity system, this appendix offers a 'talked-through' account of building one of the models in Chapter 4: the model in Figure 4.11 (Case 4). This model is relevant to exploring the research and development activity in a large manufacturing company; it was part of an SSM investigation concerned with providing information support to such researchers.

Activity Prior to Model Building

First we need to define and think about the concept to be modelled, guided by the framework which SSM provides to help model building: primary task (PT)/issue-based (IB), Root Definition, PQR, CATWOE, E_1, E_2, E_3.

For a large science-based company with international markets and competitors, the concept of providing information support to research and development is obviously highly relevant, especially given SSM's concept of the nature of information systems – as described at the start of Chapter 4. The client group for the investigation was the information department which provided that support.

The concept underpinning research and development (R&D) in the company was that there should always be a sponsor for a research or development project, people who cared about the outcome of the research. This was not 'blue-sky' research. So, in the concept being modelled, there would always be a link between sponsors and researchers which had to be maintained. Therefore this would be an activity across departments, hence an Issue-based rather than a Primary-task definition. Also, the broader concept underlying research and development was that in a science-based global business continuous R&D was essential to maintain the long-term viability of the company.

These considerations give us the following definitions for PQR, Root Definition, CATWOE, and E_1, E_2, E_3, which are relevant to carrying out R&D in this company.

PQR

P Maintain and develop a knowledge base in science and technology within the corporation
Q by defining and carrying out R&D in a sponsor/researcher relationship
R contribute to maintaining good company performance and viability.

Root Definition

A company-owned system, staffed by skilled professionals, which, in the company context, via a sponsor/researcher relationship, maintains and develops an appropriate scientific and technological base in the company by carrying out R&D in order to contribute to the long-term viability of its business.

CATWOE

C Senior management in the company
A Skilled professionals (as sponsors and researchers)

T Carry out R&D via a sponsor/researcher relationship
W R&D, continuously carried out in a science-based business can contribute to company performance and viability
O Senior company management
E Company culture and norms: in summary – define and carry out work; document and report it in explicit procedures; do both research-push and market-pull R&D; sponsor/researcher relationship

E_1, E_2, E_3

E_1 (Efficacy)	Demonstrable R&D documented results continuously emerging from the sponsor/researcher combinations
E_2 (Efficiency)	Judgement by company management that the positive effect of the R&D activity is worth the investment to achieve it
E_3 (Effectiveness)	Judgement by company management that the R&D activity is helping the business; technological base at least as good as that of competitors; lack of technical surprises from competitors

Model Building

The core of the model will clearly be activities concerned with *defining* R&D projects and *carrying them out*. Given the emphasis on sponsorship, this will have to be done in the light of the relationship with the sponsor for the research; so we can make these two activities contingent upon *maintaining the relationship* with the sponsor. This gives us three basic activities whose expression can be amplified from other elements in CATWOE – such as the company culture and norms, the combination of both technology-push and market-pull research, and the need to understand the business, since this is what the R&D will ultimately affect. Also doing the research will continually affect the appreciation of the business by both researchers and sponsors. So there will be some internal feedback here. This gives us the

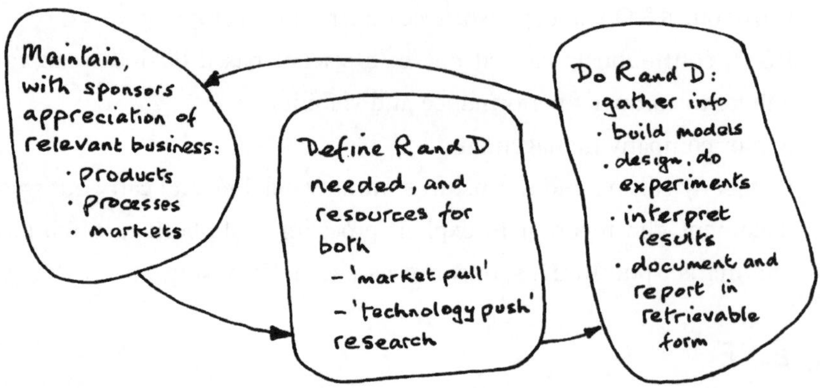

Figure A.C.1 Starting model building – basic activities deriving from the transformation process T

activities and their dependencies as shown in Figure A.C.1 (these become activities 6, 7 and 5 in the final model in Figure A.C.3.)

Next we can see that these three activities would themselves be dependent upon, and would themselves affect, the overall appreciation of the state of science and technology inside and outside the company. That, together with the research planned, would help define the skills needed, which would have to be made available. This gives three more necessary activities, yielding Figure A.C.2. (These become activities 2, 3 and 4 in the final model in Figure A.C.3.)

Finally, at a higher level, since this systems output will be judged by senior managers outside the R&D field, doing the six activities in Figure A.C.2 will all be contingent upon a good appreciation of the whole company context within which this R&D is designed and carried out. This seventh activity (activity 1) completes the operational part of the model; adding the monitoring and control then completes the model to yield Figure A.C.3.

Testing such a model consists of checking that elements in the framework are captured in the final model, which should contain the *minimum* but

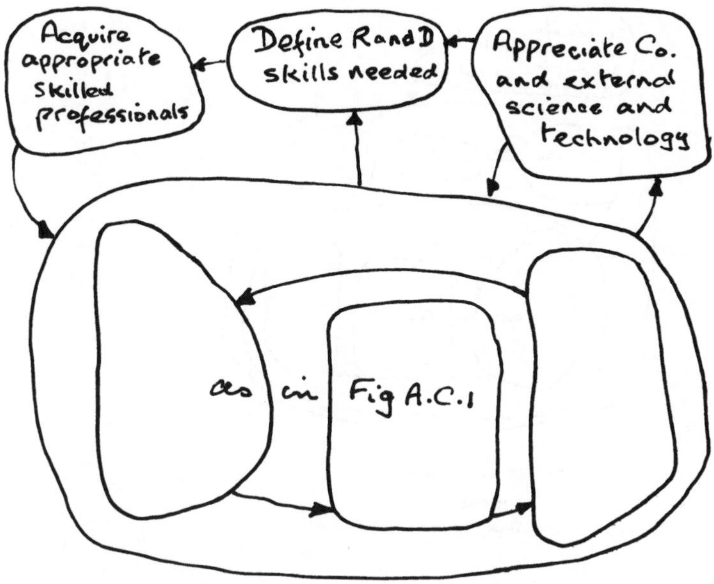

Figure A.C.2 Adding activities that T is dependent on

necessary activities to achieve this. Another useful technique is to assess the dependency of one activity upon another. Thus any activity with an arrow leaving it but none entering it should be an activity which anyone could in principle immediately do, since it is not dependent upon any other activity. This is clearly the case with activity 1 here. Similarly, for example, you could not do activity 4 unless you had already done activities 1 and 3.

So what have we created here? Like all such models in SSM it is *not* a description of R&D in this company. It is, fundamentally, only a device, a logical machine to carry out the purposeful activity described in PQR, Root Definition, CATWOE, etc. It was, in practice, an excellent source of cogent questions to ask of the real situation in order to gain insight into, and get a feel for the company context in which the information department was operating. What then came from it, technically, was another model which expanded activity 7 in greater detail. Doing this gave a model which enabled us to ask of each of its activities: What information would be needed to

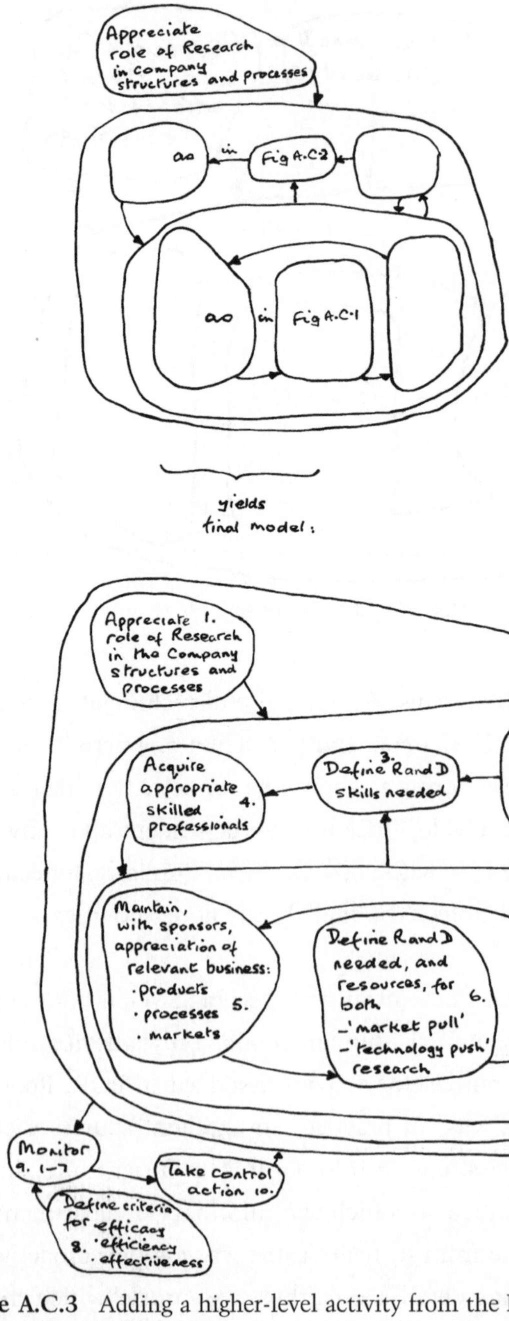

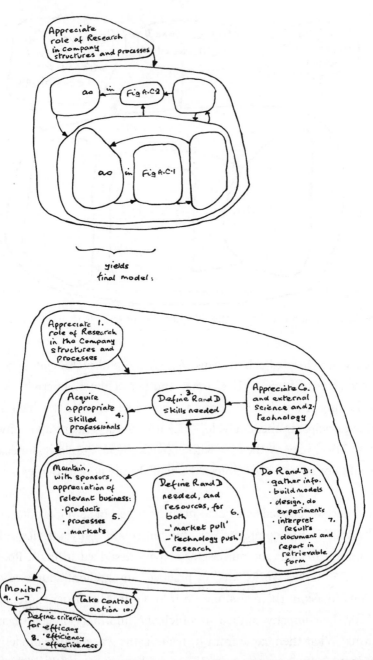

Figure A.C.3 Adding a higher-level activity from the R&D context, and the monitoring and control activities

support this activity? What is currently provided? What could the information department do to improve this? etc.

Probably the greatest difficulty in this kind of modelling, especially when it is first attempted, is to turn your back on the real situation, and focus only on the words in PQR, CATWOE, etc. – which themselves need to be thought about with great care. Practice soon overcomes this difficulty. Finally, in terms of skill, the only one needed to build useful models is a bit of clear logical thinking.

Appendix D
A Generic Model of Purposeful Activity Modelling

Chapter 2 described the framework of ideas which enable any purposeful activity model to be built (Root Definition, CATWOE, etc.), and a number of models used in SSM investigations have been presented in Chapters 3 and 4. Since the framework is always the same, it must be possible to formulate a general root definition of 'a system to build an SSM-style activity model' using the ideas in the framework. That is what is done here. Doing this illustrates that making these intellectual devices (whose purpose is to generate questions to ask of the situation addressed) is based on a clear set of internally consistent ideas.

The rather ponderous model which emerges from this exercise is not intended as a template for model construction, though we recognize that the desperate might wish to use it as such! The ungainly nature of this particular model stems from the navel-gazing aspect of it. We are here using the SSM modelling process to make a model of itself.

A Generic Model of a System to Build an SSM Activity Model

The framework to help model building in SSM is as follows:

- A written-out Root Definition based on a declared worldview
- Primary task (PT) or Issue-based (IB)? (Is there an institutionalized version or not?)
- P, Q, R (Do P, by Q, to contribute to achieving R)
- CATWOE (See Chapter 2)
- Criteria E_1, E_2, E_3 for Efficacy, Efficiency, Effectiveness (See Chapter 2)

Prior to model building, think about aspects 1 to 5:

1. Think about and express the worldview within which this model makes sense;
2. Decide whether this is a PT or IB model;
3. Define the activity to be modelled in terms of PQR;
4. Justify the P–Q link, i.e. make sure Q is an appropriate 'how' for doing the 'what' defined by P;
5. Express T (the purposeful transforming process) as changing input I into output O, making sure that the entity I is present in O but in a now-transformed state. For example, 'need for food' can be transformed into 'need for food met', *not* into 'food'. (Getting this wrong is the most common fault in SSM modelling. The point is that the fact that the need for X *leads to* X being produced is a different concept from *is transformed into*. The entity 'need' in this example is transformed into 'need met'. An abstract 'need' for food may *lead to* food being produced, but it cannot be *transformed into* plates of fish and chips!)

In this example the aspects 1 to 5 would lead to the following considerations.

(1) The worldview here embodies the view that it is both possible and useful to express the SSM modelling framework in the form of an SSM-style activity model.
(2) This modelling activity would not be institutionalized (there would not and should not be a model-building department or section, so this is an IB model).
(3) PQR would be as follows. P: build a model of the SSM modelling process by (Q) using only the SSM framework (RD, CATWOE, etc.) in order to contribute (R) to establishing that SSM modelling is coherently supported.
(4) Q is relevant to P because the framework has been used in real life to help produce hundreds of activity models. So it must be an appropriate 'how' for this activity of 'building a model'.
(5) T would be: need for a model of the SSM modelling process transformed into that need met.

These aspects cover PQR and PT/IB. The rest of the framework (CATWOE, Root Definition, E_1, E_2, E_3) would be defined as follows:

CATWOE

C The modeller(s) and those involved in the investigation
A The modeller(s) and, if possible, those involved in using the model
T Need for a model of the modelling process transformed into that need met by use of the framework
W It would be possible and useful to make such a model, showing the logical coherence of the modelling process
O The modeller(s)
E The proven framework of ideas used in SSM to help with model building

Root Definition

An activity system within the declared worldview W (above), owned and staffed (O and A) as above, which, under constraint E (above) and with beneficiaries/victims (C) as above, meets the need for a model of the SSM modelling process by using the ideas in the SSM framework, in order to contribute to showing the logical coherence of the modelling process in SSM

E_1 (Efficacy)	demonstrable model completed
E_2 (Efficiency)	an acceptable amount of time spent modelling
E_3 (Effectiveness)	agreement that this model makes clear the logic of the modelling process

The core of any SSM-style activity model will be three activities: obtain the input; transform it into the output; do something with the output. These activities, however, will not be defined in an absolute sense; how they are defined will be contingent upon other activities which have to be present because of the other CATWOE elements C, A, O, E. These, in turn will be dependent upon the pre-modelling activities 1 to 5 above and defining CATWOE, etc.).

Thus a generic model will have the general form shown in Figure A.D.1. The paradox here – that the 'pre-modelling activities' are themselves inserted into the model, so that they too are susceptible to monitoring and control activity – is not a problem. Rather it illustrates an important point, namely that the pre-modelling activities not only influence what activities will be included in the model, but also may themselves be influenced by those activities, as modelling proceeds, and may cause some rethinking. That is to say the relationship between pre-modelling activity and modelling activity is one of influence on each other; neither is prime:

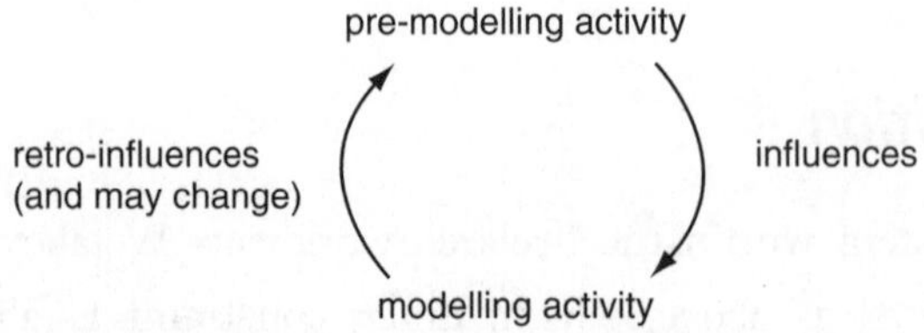

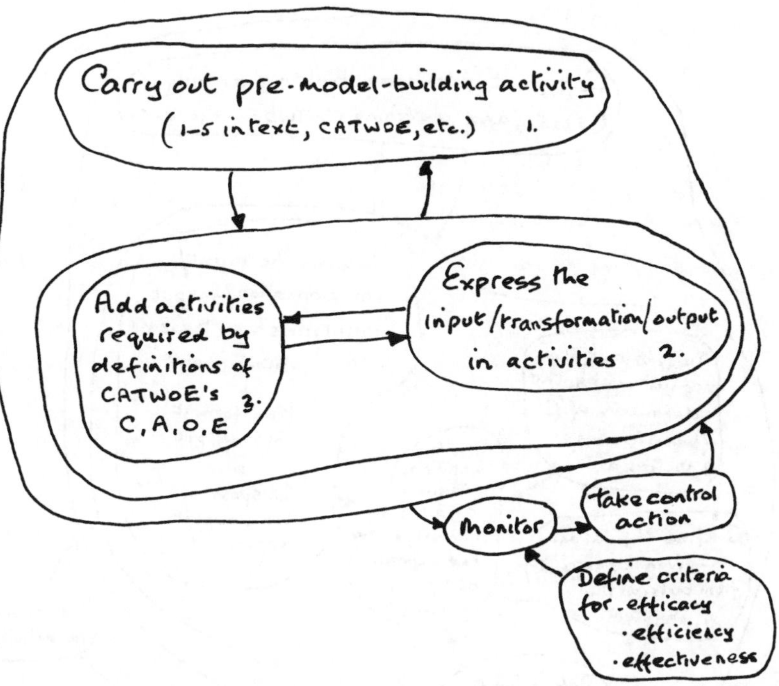

Figure A.D.1 The general form of the generic model

So model building, like most elements of SSM, is part of a process of *learning*. The model which results from these considerations is shown in Figure A.D.2. As usual the closure effected by the 'take control activity' ensures that the whole model building process is one of learning.

In the model itself, the inclusion of activity due to the O in CATWOE (the person or persons who could stop the activity – here the modeller(s)) could have been included in the operational part of the model, alongside activities 2, 3 and 4. Instead it has been included as part of the 'monitoring and control' activity. This is done here to draw attention to the fact that the activity system 'owner' (O of CATWOE) will always be strongly linked to the criteria for effectiveness (E_3), the higher-level or longer-term aim of the purposeful activity. O will stop the overall activity if it does not seem to be

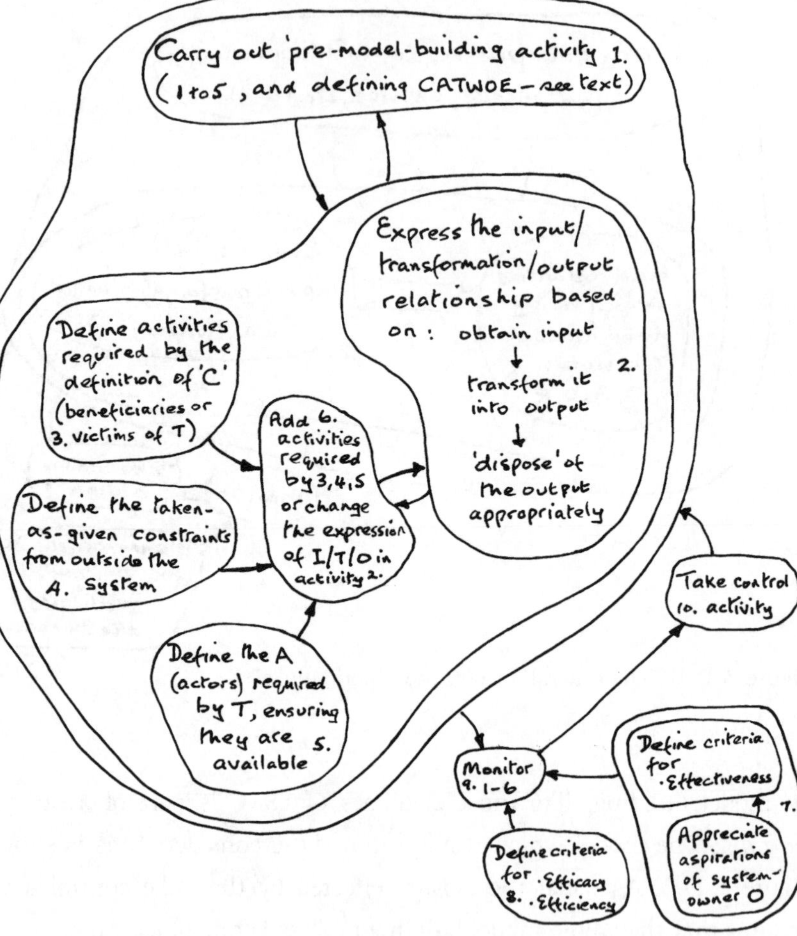

Figure A.D.2 The generic model of the modelling process

contributing to that higher level aim. Thus, in the model built in Appendix C, the owner O is the company as represented by its senior management. They could change or close down the expensive research and development activity if in their judgement it was not contributing to long-term viability.

Appendix E
Some Frequently Asked Questions

Does SSM have rules?

Since it is a methodology (a set of principles which underpin the action undertaken in a specific situation) the idea of 'rules' is not a very helpful concept. SSM's basic principle is to think separately about, and then relate, two different elements: perceptions people have of a complex human situation, and constructed conceptualizations aimed at gaining insights into that situation. These elements are related by using the conceptualizations to define good questions which are then asked about the multiple perceptions of the situation, thus providing coherent structure to discussion (and also generating new conceptualizations in a cyclic process of learning). This leads to action to improve the situation by finding possible change which is both desirable and feasible. Such rules as there are within SSM concern the tools used to create the conceptualizations. For example, every element in a model of purposeful activity must be based on a *verb* in the imperative. For example: '*Obtain* raw material x.' Avoid expressing the elements as entities (such as 'Raw material x').

What kind of thinking processes are used in SSM?

Both the logical/analytical mode of thinking (as in model building) and the slower sense-making mode – as you allow the situation to 'speak to you'. The 'finding out' phase in SSM calls for empathizing, and a conscious holding back from quickly reaching conclusions or (especially) apparent 'solutions' which may become dogma and close down broader thought.

How does SSM identify ideas for improvement?

It doesn't. It enables and helps those using SSM to make sense of the problematical situation and to learn their way to identifying well-thought-out ideas for improvement. SSM brings gentle open-handed structuring to the process of inquiry, which helps anyone to make best use of their intellectual capabilities.

Is there a wrong way to use SSM?

We would baulk at the use of the word 'wrong', but using SSM prescriptively, as if it were a recipe to be followed slavishly, throws away much of its value as methodology. Proper use of SSM always entails getting a 'feel' for a situation. The danger of a prescriptive approach is that it may quickly force perceptions into a particular shape.

When should SSM be used?

It can be used whenever, in human situations, the feeling arises that 'this could/should be improved', or 'something needs to be done about this', or 'I feel uneasy about this, it needs looking at'. It can also be used – especially to help understand the context of situations and underlying issues – whenever a project is mounted to achieve some apparently desired and defined end. On the other hand, once internalized as a natural way of thinking, it can be used to guide the process of 'managing' anything.

How long does an SSM investigation take?

The ideas are not tied to any particular timescale. Some studies have taken an hour or so, some a year or two. In the former case the 'finding out' has to take the existing knowledge level as given, which may of course be limiting. Parts of SSM can be used instantly. If you are sitting in a meeting and the chairperson proposes that some particular action should be taken, you can usefully, in your head, make a CATWOE and PQR analysis of it and at once pose some cogent questions.

How does SSM compare with or relate to approaches such as SSADM, IDEF, PRINCE, etc.?

It is not in the same game. It is much less prescriptive and takes much less as given, compared with these methods. They provide a means of planning and documenting a sequence of actions aimed at achieving a known end, such as, for example, software writing. Hence they may be relevant towards the end of an SSM investigation, when 'what to do' has been decided. SSM addresses the problems to be overcome in deciding what the desirable end or objective is. It would normally be used before more prescriptive toolsets such as SSADM or IDEF or PRINCE are called into play. The UK Government Central Computer and Telecommunication Agency, which selected SSADM as a required procedure within government departments, published documents suggesting that SSADM's early stages could be enhanced by a prior study based on SSM (e.g. *Applying Soft Systems Methodology to an SSADM Feasibility Study*, HMSO, 1993). Similarly SSM can build up understanding of a spectrum of issues before the detailed planning and conduct of a project is set out in PRINCE or PRINCE 2.

An account of SSM's role in project management in general is given in a paper by Mark Winter and Peter Checkland: Soft systems: a fresh perspective for project management, in *Civil Engineering*, Vol 156(4), 187–192, 2003.

Why is the literature of SSM full of hand-drawn diagrams?

It's a matter of psychology. The literatures of control engineering and management science have many diagrams dominated by straight lines, right angles and rectangular boxes. These convey the impression: this is the case, full stop! The hand-drawn diagrams in the SSM literature aim to convey an organic rather than a mechanical impression. They underline that absolute certainty is forever elusive in human affairs; they are working diagrams, part of the learning process. And they look more human, more attractive than straight lines and right angles.

Index